C000192918

ISBN 978-0-267-63607-5
PIBN 10053177

THE

PHILOSOPHY OF HEALTH;

OR,

AN EXPOSITION

OF THE

PHYSICAL AND MENTAL CONSTITUTION OF MAN,

WITH A VIEW TO THE PROMOTION OF

HUMAN LONGEVITY AND HAPPINESS.

BY

SOUTHWOOD SMITH, M.D.,

Physician to the London Fever Hospital, to the Eastern Dispensary, and to the Jews' Hospital.

IN TWO VOLUMES. VOL. II.

THIRD EDITION.

LONDON:

C. COX, 12, KING WILLIAM STREET, STRAND.

1847.

CONTENTS OF VOL. II.

CHAPTER VIII.

OF THE FUNCTION OF RESPIRATION.

CHAPTER IX.

OF THE FUNCTION OF GENERATING HEAT.

CHAPTER X.

OF THE FUNCTION OF DIGESTION.

CHAPTER XI.

OF THE FUNCTION OF SECRETION.

CHAPTER XII.

OF THE FUNCTION OF ABSORPTION.

CHAPTER XIII.

OF THE FUNCTION OF EXCRETION.

CHAPTER XIV.

OF THE FUNCTION OF NUTRITION.

THE

PHILOSOPHY OF HEALTH.

CHAPTER VIII.

OF RESPIRATION.

Respiration in the plant; in the animal—Aquatic and aerial respiration—Apparatus of each traced through the lower to the higher classes of animals—Apparatus in man—Trachea, Bronchi, Air Vesicles—Pulmonary artery—Lung—Respiratory motions: inspiration; expiration—How in the former air and blood flow to the lung; how in the latter air and blood flow from the lung—Relation between respiration and circulation—Quantity of air and blood employed in each respiratory action—Calculations founded on these estimates—Changes produced by animal respiration on the air: changes produced by vegetable respiration on the air—Changes produced by respiration on the blood—Respiratory function of the liver—Uses of respiration.

313. No organized being can live without food and no food can nourish without air. In all creatures the necessity for air is more urgent than that for food, for some can live days, and even weeks, without a fresh supply of food, but none without a constant renewal of the air.

314. The food having undergone the requisite preparation in the apparatus provided for its assimilation, is brought into contact with the air, from which it abstracts certain principles, and to which it gives others in return. By this interchange of principles the composition of the food is changed: it acquires the qualities necessary for its combination with the living body. The process by which the air is brought into contact with the food, and by which the food receives from the air the qualities which fit it for becoming a constituent part of the living body, constitutes the function of respiration.

315. In the plant, the air and the food meet in contact and react on each other in the leaf. The crude food of the plant having in its ascent from the root through the stalk, received successive additions of organic substances, by which its nature is assimilated to the chemical condition of the proper nutritive fluid of the plant (320 and 325), undergoes in the leaf a double process; that of Digestion and that of Respiration. The upper surface of the leaf is a digestive apparatus, analogous to the stomach of the animal; the under surface of the leaf is a respiratory apparatus, analogous to the lung of the animal. For the performance of this double function, incessantly carried on by the leaf, its organization is admirably adapted.

316. The solid skeleton of the leaf consists of a net-work composed partly of woody fibres and partly of spiral vessels which proceed from the stem, and which are called veins (fig. CXXII. 1, 3).

Fig. CXXII.

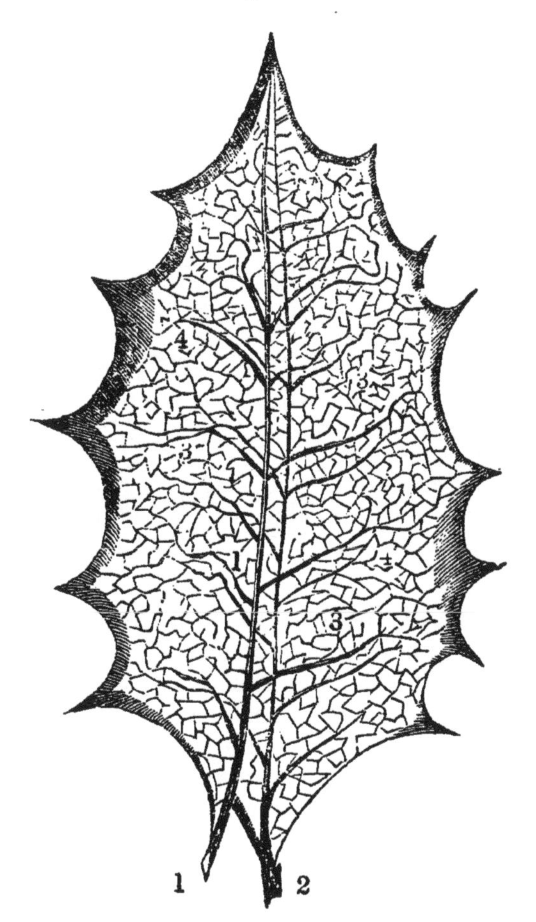

View of the net-work which forms the solid structure of the leaf, and which consists partly of woody fibres, and partly of spiral vessels. 1. Vessels of the upper surface; 2. vessels of the under surface; 3. distribution of the vessels through the substance of the leaf; 4. interspaces between the vessels occupied by parenchyma or cellular tissue.

In the interstices between the veins is disposed a quantity of cellular tissue, termed the parenchyma of the leaf (fig. CXXII. 4): the whole is enveloped in a membrane, called the cuticle (fig. CXXIII. 1),

Fig. CXXIII.

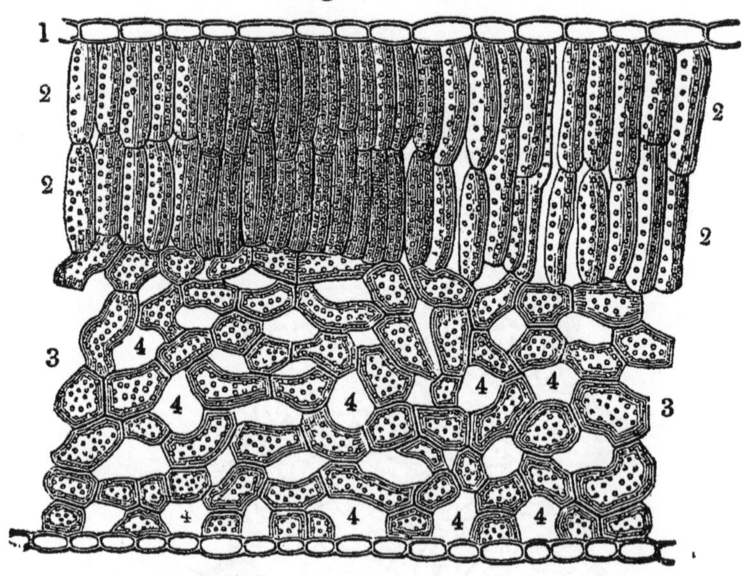

Vertical section of the leaf as it appears when seen highly magnified under the microscope. 1. Cells of the cuticle filled with air; 2. double series of cylindrical cells occupying the upper surface of the leaf filled with organic particles; 3. irregular cells forming a reticulated texture occupying the under surface of the leaf; 4. interspaces between the cells, termed the intercellular passages or air chambers.

which is furnished with apertures denominated stomata, or stomates (fig. CXXIV.).

317. The cuticle consists of a layer of minute cellules, colourless, transparent, without vessels, without organic particles of any kind, and probably filled with air (fig. CXXIII. 1). These cel-

lules open externally, at certain portions of the cuticle, by apertures or passages which constitute the stomates (fig. CXXIV.), and which present the

Fig. CXXIV.

View of the stomata of a leaf, some of them represented as open and others as closed.

appearance of areolæ with a slit in the centre (fig. CXXIV.). They form a kind of oval sphincters, which are capable of opening or shutting, according to circumstances, and they are disposed on both surfaces of the leaf, but most abundantly on the under surface, excepting in leaves which float on water, in which they are always on the upper surface only.

318. The cellular tissue or parenchyma, immediately beneath the cuticle, when examined in thin slices, and viewed under a microscope with a high magnifying power, presents a regular structure disposed in perfect order. It consists, on the upper surface, of a layer, and sometimes of two and even three layers, of vesicles of an oblong or

cylindrical form, placed perpendicularly to the surface of the leaf, set close to each other (fig. cxxiii. 2), and filled with organic particles constituting the green matter which determines the colour of the leaf. On the under surface, on the contrary, the vesicles, which are larger than the cylindrical, are of an irregular figure, and are placed in an horizontal direction, at such distances as to leave wide intervals between each other (fig. cxxiii. 3); yet uniting and anastomosing together, and thus forming a reticulated tissue, presenting the appearance of a net with large meshes (fig. cxxiii. 3).

319. A leaf, then, consists of a double congeries of vesicles containing organic particles, penetrated by woody fibre and air vessels (which is probably the true nature of the spiral vessels), the whole being enclosed within a hollow stratum of air-cells.

320. The crude sap, composed principally of water, holding in solution carbonic acid, acetic acid, sugar, and a matter analogous to gum, is transmitted through the leaf-stalk to the cylindrical vesicles of the upper surface of the leaf (fig. cxxiii. 2). These vesicles exhale a large proportion of the water; the evaporation of which is so powerfully assisted by the action of the sun's rays, that it would probably become excessive, were it not for the perpendicular direction of the cylindrical vesicles (fig. cxxiii. 2); but in consequence of their being disposed perpendicularly to

the surface of the leaf, their ends only are presented towards the heavens (fig. CXXIII. 2), and thus the main part of their surface is protected from the direct influence of the solar rays. The primary effect of the evaporation carried on in the cylindrical vesicles, is the condensation of the organic matters contained in the sap.

321. At the same time that the cylindrical vesicles pour the superfluous water of the sap into the surrounding atmosphere, they abstract from the atmosphere in return carbonic acid, which, together with that already contained in the sap, is decomposed. The oxygen is evolved; the carbon is retained. The physical agent by which this chemical change, which constitutes the digestive process of the plant, is effected, is the solar ray; hence the vesicles which contain the fluid to be decomposed, are placed on the upper surface of the leaf, where their contents are fully exposed to the action of the sun; and hence also this process takes place only during the day, and most powerfully under the direct solar ray: but although the direct influence of the sun be highly conducive to the process, yet it is not indispensable to it; for it goes on in daylight although there be no sunshine. Light, then, would appear to be the physical agent which effects on the crude food of the plant a change analogous to that produced on the crude food of the animal by the juices of the stomach.

322. After the sap has been elaborated in the cylindrical vesicles, by the exhalation of its watery particles, by the condensation of its organic matter, by the retention of carbon and the evolution of oxygen, it is transmitted to the reticulated vesicles of the under surface of the leaf (fig. CXXIII. 3). These vesicles, large, loose, and expanded, as they have an opposite function to perform, are arranged in a mode the very reverse of the cylindrical: in such a manner as to present the greatest possible extent of surface to the surrounding air (fig. CXXIII. 3): at the same time the broad interspaces between them (fig. CXXIII. 4) are so many cavernous air-chambers into which the air is admitted through the stomates (fig. CXXIV.). The cylindrical vesicles, exposed to the direct rays of the sun, are protected by the closeness with which they are packed; and by the small extent of surface they present to the heavens: the reticulated vesicles, whose function requires that they should have the freest possible exposure to the surrounding air, are protected from the solar ray, first by their position on the under surface of the leaf; and, secondly, by the dense and thick barrier formed by the stratum of cylindrical vesicles (fig. CXXIII. 2).

323. In the cylindrical vesicles carbonic acid is decomposed; in the reticulated vesicles, on the contrary, carbonic acid is re-formed. The oxygen required for this generation of carbonic acid is abstracted partly from the surrounding air; the

carbon is derived partly, perhaps, from the air, but chiefly from the digested sap, and the carbonic acid, formed by the union of these elements, is evolved into the surrounding atmosphere.

324. This operation, which is strictly analogous to that of respiration in the animal, in which carbonic acid is always generated and expired, is carried on chiefly in the night. In this manner, under the influence of the solar light, the leaf decomposes carbonic acid; retains the carbon and returns the greater part of the oxygen to the air in a gaseous form. At night, in the absence of the solar ray, the leaf absorbs oxygen, combines this oxygen with the materials of the sap to produce carbonic acid, which, as soon as formed, is evolved into the surrounding air. The carbonic acid gas exhaled during the night is re-absorbed during the day and oxygen is evolved; and this alternate action goes on without ceasing; whence the plant deteriorates the air by night, by the abstraction of its oxygen and the exhalation of carbonic acid; and purifies it by day by the evolution of oxygen and the abstraction of carbonic acid.

325. The result of these chemical actions is the conversion of the crude sap into the proper nutritive juice of the plant. When it reaches the cylindrical vesicles, the sap is colourless, not coagulable, without globules, composed chiefly of water holding in solution carbonic and acetic acids, sugar, gum, and several salts; when it leaves

the reticulated vesicles it is a greenish fluid, partly coagulable and abounding with organic particles under the form of globules. Its chemical composition is now wholly changed; it consists of resinous matter, starch, gluten, and vegetable albumen. It is now thoroughly elaborated nutritive fluid; the proper food of the plant (cambium); rich in all the principles which are fitted to form vegetable secretions: it is to the plant what arterial blood is to the animal, and like the vital fluid formed in the lung, the cambium elaborated in the leaf, is transmitted to the different parts and organs of the plant to serve for their nutrition and development.

326. The formation of this nutritive fluid by the plant is a vital process, as necessary to the continuance of its existence, as the process of sanguification is necessary to the maintenance of the life of the animal. If the plant be deprived of its leaves, if the cold destroy, or the insect devour them, the nutrition of the plant is arrested; the development of the flowers, the maturation of the fruit, the fecundation of the seeds, all are stopped at once, and the plant itself perishes.

327. The proper nutritive juice of the plant, completed by the process of respiration, is formed by the elaboration of organic combinations of a higher nature than those afforded by the sap. Acid, sugar, gum (325) are converted into the higher organic compounds, resin, gluten, starch,

albumen, probably by chemical processes, the result of which is the inversion of the relative proportions of oxygen and carbon. In the organic matters contained in the sap, the proportion of oxygen, compared with that of carbon, is in excess; on the contrary, in the higher compounds contained in the cambium, the carbon preponderates: by the inversion of the relative proportions of these two elements, the organic compounds of a lower nature, appear to be changed into those of a higher; to be brought into a chemical condition nearer to that of the proper substance of the plant; a condition in which they receive the last degree of elaboration preparatory to their conversion into that substance.

328. In the process of respiration in the animal, as in the plant, parts of the digested aliment mix with the air; parts of the air mix with the digested aliment; and by this interchange of principles, the chemical composition of the aliment acquires the closest affinity to that of the animal body; is rendered fit to combine with it; fit to become a constituent part of it.

329. The extent and complexity of the respiratory apparatus in the animal, is in the direct ratio of the elevation of its structure and the activity of its function, to which the quantity of air consumed by it is always strictly proportionate.

330. The process of respiration in the animal is effected by two media, air and water; but the

only real agent is the air; for the water contributes to the function only by the air contained in it. Respiration by water is termed aquatic, that by the atmosphere, atmospheric or aërial respiration.

331 The quantity of air contained in water being small, aquatic is proportionally less energetic than aërial respiration; and, accordingly, the creatures placed at the bottom of the animal scale, having the simplest structure and the narrowest range of function, are all aquatic.

332. Whatever the medium breathed, respiration in the animal is energetic in proportion to the extent of the respiratory surface exposed to the surrounding element. As the water-breathing animals successively rise in organization, their respiratory surface becomes more and more extended, and a proportionally larger quantity of water is made to flow over it. It is the same in aërial respiration: the higher the animal, the greater the extent of its respiratory surface; and the larger the bulk of air that acts upon it.

333. Whatever the medium breathed, respiration is effected by the contact of fresh strata of the surrounding element with the respiratory surface. The mode in which this constant renewal of the strata is effected, is either by the motion of the body to and fro in the element; or by the creation of currents in it, which flow to the respiratory surface. A main part of the apparatus of

respiration consists of the expedients necessary to accomplish these two objects; and that apparatus is simple, or complex, chiefly according to the extent of the mechanism requisite to effect them.

334. Whatever the medium breathed, the organic tissue which constitutes the essential part of the immediate organ of respiration is the skin. The primary tissue of which the skin is composed is the cellular (23 et seq.), which, organized into mucous membrane (33 et seq.), forms the essential constituent of the skin (34). In all animals the skin covers both the external and the internal surfaces of the body (34). When forming the external envelop, this organ commonly retains the name of skin; when forming the internal lining, it is generally called mucous membrane; and in all animals, from the monad to man, either in the form of an external envelop, or an internal lining, or by both in conjunction, or by some localization and modification of both, the skin constitutes the immediate organ of respiration. In different classes of animals it is variously arranged, assumes various forms, and is placed in various situations, according to the medium breathed, and the facility of bringing its entire surface into contact with the surrounding element; but in all, the organ and its office are the same: it is the modification only —that modification being invariably and strictly adaptation, which constitutes the whole diversity of the immediate organ of respiration.

335. At the commencement of the animal scale, in the countless tribes of the polygastrica (vol. i. p. 34, et seq.), respiration is effected through the delicate membrane which envelops the soft substance of which their body is composed. The air contained in the water in which they live, penetrating the porous external envelop, permeates every part of their body; aërates their nutritive juices; and converts them immediately into the very substance of their body. They are not yet covered with solid shells, nor with dense impervious scales, nor with any hard material which would exclude the general respiratory influence of water, or render necessary any special expedient to bring their respiratory surface into contact with the element.

336. But in some tribes even of these simple creatures there is visible by the microscope an afflux of their nutritive juices to the delicate pellicle that envelops them, in the form of a vascular net-work, in which there appears to be a motion of fluids, probably the nutritive juices flowing in the only position of the body in which they could come into direct contact with the surrounding element. In some more highly advanced tribes, as in wheel animalcules, there is an obvious circulating system in vessels near the surface of the skin. In other tribes, the internal surface constituting the alimentary canal, is of great extent and width, and forms numerous cavities which

are often distended with water. In this manner a portion of the internal, as well as the external surface is made contributary to the function of respiration, and this extended respiration is conducive to their great and continued activity, to their rapid development, and to the extraordinary fertility of their races.

337. In creatures somewhat higher in the scale, a portion of the external surface is reflected inwards in the form of a sac, with an external opening (fig. cxxv. 1). In some medusæ there are

Fig. CXXV.—*Medusa.*

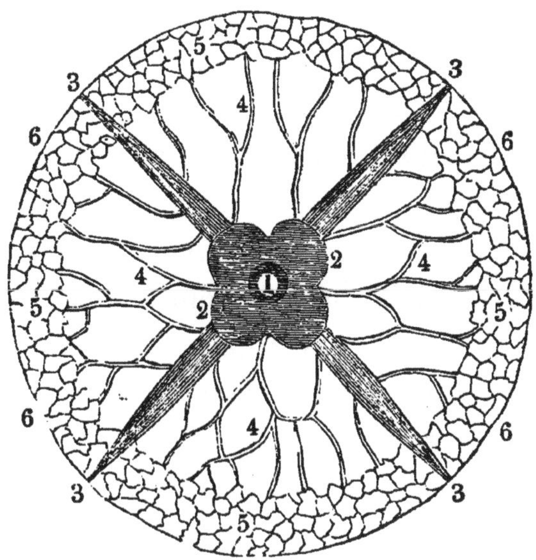

1. The mouth; 2. the stomach; 3. large canals going from the stomach; 4. smaller canals which form, 5. a plexus of vessels at the margin of the disc serving for respiration; 6. margin of the disc.

numerous sacs of this kind, which pass inwards until they are separated only by thin septa from the cavities of the stomach. The water permeating and filling these sacs comes into contact with an interior portion of the body, not to be reached through the external surface. At the margin of the disk (fig. cxxv. 6) there is spread out a delicate net-work of vessels (fig. cxxv. 5); these vessels communicate with small canals (fig. cxxv. 4) which open into larger canals (fig. cxxv. 3) that proceed directly from the stomach (fig. cxxv. 2). As the aliment is prepared by the stomach, it is transmitted thence by these communicating canals to the exterior net-work of vessels where it is aërated.

338. As organization advances, as the component tissues of the body become more dense, and are moulded into more complex structures, when, moreover, these structures are placed deep in the interior of the body, far from the external envelop, and proportionally distant from the surrounding element, the respiratory apparatus necessarily increases in complexity. The first complication consists in the formation of minute, delicate, transparent tubes (fig. cxxvi. 5), which communicate with the external surface by a special organ (fig. cxxvi. 4) that conveys water into the interior of the body (fig. cxxvi. 5). By means of these ramifying water-tubes, upon the delicate walls of which the blood-vessels are spread out in minute and beautiful capillaries, the water is brought into immediate contact with the vascular system.

Fig. CXXVI.—*Holothuria.*

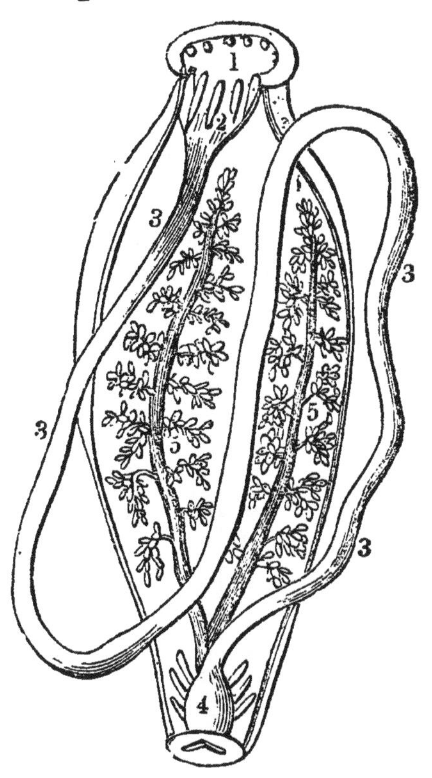

1. Mouth; 2. salivary sacs; 3. intestine; 4. cloaca; 5. ramified tubes, conveying water for respiration into the interior of the body.

339. Next, in the ascending scale, the external envelop of the body is extended into a distinct additional or supplemental organ, by which the function of the skin is assisted. This additional organ is called branchia or gill. The simplest form of branchia consists of folds or duplicatures of skin, forming ramified tufts (fig. CXXVII. 1), which in general have a regular and often a symmetrical disposition on the external surface (fig. CXXVII. 1). Sometimes, as in the water breath-

Fig. CXXVII.—*Lumbricus Marinus.*

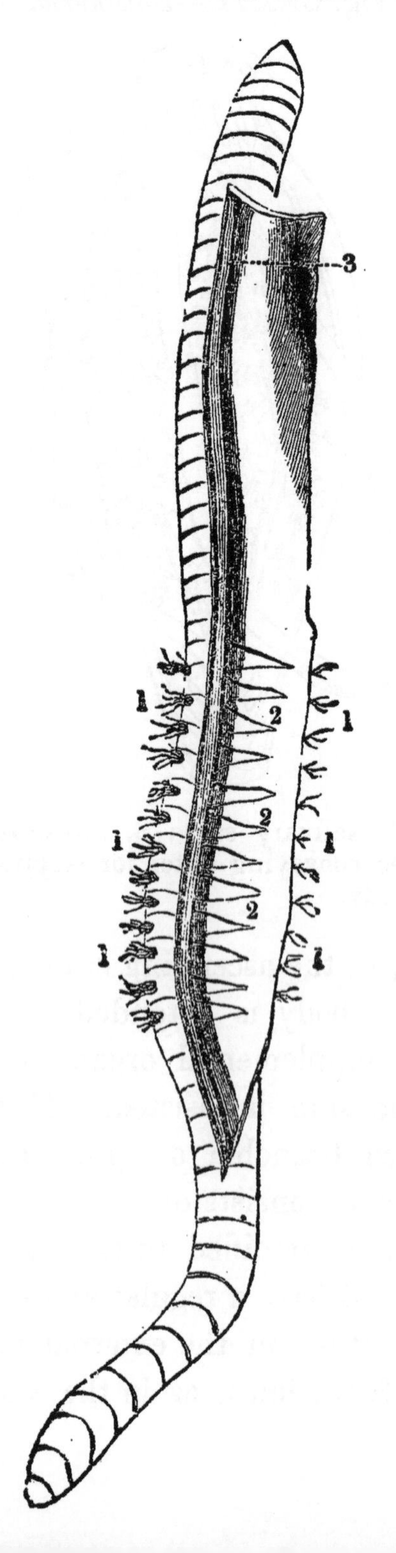

1. Respiratory tufts. 2. Artery and vein, supplying the respiratory apparatus. 3. Dorsal vessel.

ing annelides, these tufts form a fan-like expansion around the head; but at other times they are disposed in regular series along the whole extent of the body.

340. Instead of branchiæ in the form of ramified tufts, the ascending series of animals, namely, the higher crustacea, possess branchiæ composed of numerous, delicate, thin laminæ or leaves, divided from each other, yet placed in close proximity, like the teeth of a fine comb, whence this arrangement is termed pectinated. Over the blood-vessels of the system spread out on these delicate, fringed, pectinated leaves, the water is driven in constant streams.

341. Still higher in the scale, as in molluscous animals, an internal sac is formed to which are sometimes attached numerous tufts; but which at other times is itself plaited into beautifully disposed regular folds, crowded with blood-vessels and constantly bathed with fresh currents of water.

342. In all these water-breathing creatures, respiration is effected, either by the progressive motion of the body through the water, or by the creation of currents which bring fresh strata of the fluid into contact with the respiratory surfaces. Both objects are effected by the same in-

struments, namely, minute fibres having the ap-pearance of fine hairs or bristles. These fibres which are called cilia, have in general an elon-gated, flattened, thin, and tapering form (fig. cxxviii). Their number, position, and arrange-ment, are infinitely various. Sometimes, as in the poriferous animals, they are so minute that they cannot be rendered visible to the eye even by the microscope, although the evidence of their exist-ence and action is indubitable. Sometimes they

Fig. CXXVIII.

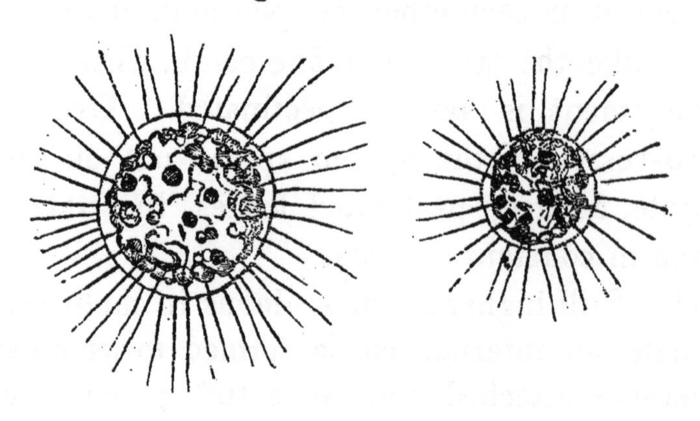

Trichoda showing the form and a frequent arrangement of Cilia.

are of great size and strength, attached by distinct ligaments to the body and moved by powerful muscles, as in wheel animalcules. Sometimes, as in polypiferous animals, they are disposed around the orifice of the polypes or upon the sides of the tentacula, the instruments by which the animal seizes its prey. Sometimes they are sym-metrically disposed in longitudinal series along the surface of the body, as in the Beroe pileus; at other times they are arranged in circles; whenever there are branchiæ, they are disposed around the margin of the branchial apertures, and always on the margins of the minute meshes which compose the branchiæ themselves.

343. In some cases the number of these cilia is immense. Each polype, for example, has usually twenty-two tentacula, and there are about fifty cilia on each side of a tentaculum, making two thousand two hundred cilia on each polype. As there are about one thousand eight hundred cells in each square inch of surface, and the branches of an ordinary specimen present about ten square inches of surface, we may estimate that an ordinary specimen of this zoophite presents more than eighteen thousand polypes, three hundred and ninety-six thousand tentacula, and thirty-nine million six hundred thousand cilia. But other species contain more than ten times these numbers. Dr. Grant has calculated that there are about four hundred million cilia on a single Flustra foliacea.

344. The motions of these cilia are regular, incessant, and when in full activity far too rapid to be distinguished by the eye even when assisted by the microscope. They are generally to be perceived only when their motions are comparatively feeble. They produce two effects. In animals capable of progressive motion, they transport the

body through the water, while they constantly bring new strata of water into contact with the respiratory surface. In this case they are partly organs of locomotion, and partly organs subservient to respiration. On the other hand, in animals which are not capable of moving from place to place, they create currents by which the respiratory surface is constantly bathed with fresh streams of water. These currents are regular, constant, unceasing. Like some physical phenomena not depending on vitality, it is a continued stream as regular as the motions of rivers from their source to the ocean, or any other movements depending on the established order of things. Dr. Grant, to whom we are indebted for our knowledge of the true nature of these currents, as well as of the instruments by which they are effected, gives the following account of the observation which led to the discovery:—"I put," says he, "a small branch of the spongia coalita, with some sea water into a watch-glass, under the microscope, and on reflecting the light of a candle through the fluid, I soon perceived that there was some intestine motion in the opaque particles floating through the water. On moving the watch-glass, so as to bring one of the apertures on the side of the sponge fully into view, I beheld, for the first time, the splendid spectacle of this living fountain, vomiting forth from a circular cavity an impetuous torrent of liquid matter, and hurling along in rapid succession opaque masses which it strewed everywhere around. The beauty and novelty of such a scene in the animal kingdom long arrested my attention, but after twenty-five minutes of constant observation, I was obliged to withdraw my eye from fatigue, without having seen the torrent for one instant change its direction, or diminish in the slightest degree the rapidity of its course. I continued to watch the same orifice, at short intervals, for five hours, sometimes observing it for a quarter of an hour at a time, but still the stream rolled on with a constant and equal velocity."

345. The simple expedients which have been described suffice for carrying on the function of respiration in the water-breathing invertebrata; but in creatures that possess a vertebral column, and the more perfect skeleton of which it forms a part, there is a prodigious advancement in the organization of the whole body, of the nervous and muscular systems especially, the organs of the animal, as well as in all the organs of the organic life. A corresponding development of the function of respiration is indispensable. Accordingly, a sudden and great development in the apparatus of this function is strikingly apparent in fishes, the lowest order of the vertebrata, in which the branchiæ, though still preserving the same form as in the animals below them, are large and complex organs. The branchiæ of fishes still

consist of fringed folds of membrane disposed, as in the preceding classes, in laminæ or leaves (fig. cxxix. 5); but there are now commonly four series of these leaves, on each side of the body, placed in close approximation to each other, the several

Fig. CXXIX.—*Diagram of the Apparatus of the Circulation and Respiration in the Fish.*

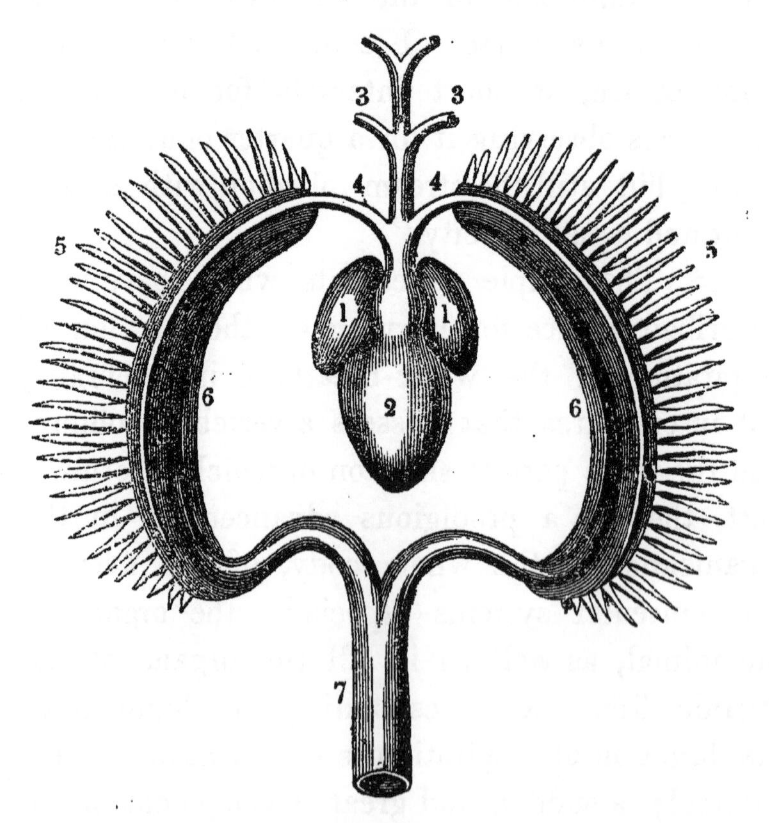

1. Auricle (Single) of the heart. 2. Ventricle (single) of the heart. 3. Trunk of the branchial artery. 4. Division of the branchial artery going to the branchiæ or gills. 5. Leaves of the branchiæ. 6. Branchial veins, which return the blood from the branchiæ, and unite to form. 7. the aorta, by the division of which the aërated blood is carried out to the system.

leaves being divided into minute fibres, which are set close like the barbs of a feather, or the teeth of a fine comb (fig. CXXIX. 5). Each leaf rests either on a cartilaginous or a bony arch, which exactly resembles the rib of the more perfect skeleton, and performs a strictly analogous function; for these arches are capable of alternately separating from, and of approximating to, each other, and these alternate motions are effected by appropriate muscles. As these movements of separation or approximation take place, the branchiæ are either opened or closed, and their surface proportionally expanded or contracted. Upon these leaves (fig. CXXIX. 5) the veins (347) of the system (fig. CXXIX. 4) are spread out in a state of capillary division of extreme minuteness, forming a net-work of vessels of extreme tenuity and delicacy. So prodigiously is the surface increased for the expansion of these vessels by the leaf-like disposition of the branchiæ, that it is computed that the branchial surface of the skate is at least equal to the surface of the whole human body.

346. Through this extended surface the whole blood of the system must circulate, and every point of it must be unceasingly bathed with fresh streams of water. To generate the force necessary for the accomplishment of these objects, an increase of power must be communicated both to the circulating and to the respiratory apparatus. Neither the contractile power of the vessels by

which in some of the simpler animals the nutritive fluid is put in motion, nor the contraction of the rudimentary heart by which in creatures somewhat higher in the scale a more decided impulse is given to the blood, are sufficient. A muscular heart, capable of acting with great power, is now constructed, which is placed in such a position as to enable it to propel with velocity the whole blood of the body through the myriads of capillary vessels that crowd every point of the surface of the branchial leaflets. To bring the water with the requisite degree of force into contact with this flowing stream, the apparatus of cilia is wholly inadequate. The water entering by the mouth, is driven with force, by the powerful muscles of the thorax, through apertures that lead to the branchial cavities. At the instant that the branchial leaves receive the currents of water through the appropriate apertures, the cartilaginous or bony arches which sustain the leaves, separate to some distance from each other, and to that extent expand the leaves and proportionally increase the surface exposed to the water: at the same time, the rush of water through the leaves unfolds and separates each of the thousand minute filaments of which they are composed, so that they all receive the full action of the fluid as it flows over them.

347. After the venous blood of the system has been thus exposed to the action of the respiratory

medium, it is taken up by the vessels called the branchial veins (fig. CXXIX. 6), which for the reason assigned (372) are functionally arteries, as the branchial artery (fig. CXXIX. 4) is functionally a vein. The branchial veins uniting together form the great arterial trunk of the system, (fig. CXXIX. 7) by which the aërated blood is carried out to every part of the body.

348. But as if even this extent of apparatus were insufficient to afford the amount of respiration required by the system of the fish, the entire surface of its body, which in general is naked and highly vascular, respires like the branchiæ. Moreover, many fishes swallow large draughts of air, by which they aërate the mucous surface of their alimentary canal, which also is highly vascular; and still further, numerous tribes of these animals are provided with a distinct additional organ, a bag placed along the middle of the back filled with air. Commonly this air bag communicates with some part of the alimentary canal near the stomach, by means of a short wide canal termed the ductus pneumaticus, but sometimes it forms a simple shut sac without any manifest opening; at other times it is divided and subdivided in a perfectly regular manner, forming extended ramified tubes; while at other times its ramifications present the appearance of so many pulmonary cells. It is the rudiment of the complex lung of the higher vertebrata, and it assists respiration;

although since in some tribes it contains not atmospheric air but azote, it is without doubt subservient to other uses in the economy of the animal.

349. In water-breathing animals, from the lowest to the highest, it is then manifest that a special apparatus is provided for, constantly renewing the streams of water that are brought into contact with their respiratory surface.

350. It is the same in aërial respiration. In the simplest form of aërial respiration the apparatus consists of minute bags or sacs, placed commonly in pairs along the back, which open for the admission of the air on the external surface, by small orifices called spiracula or spiracles

Fig. CXXX.—*Tracheæ.*

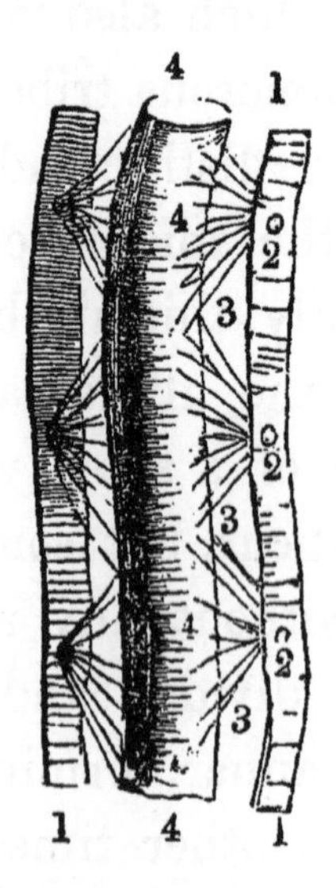

1. Integument or skin of the body. 2. Spiracula opening on the external surface of the skin. 3. Tracheæ, or air tubes, proceeding in form of radii from the spiracles to 4. the alimentary canal.

(fig. cxxx. 2), at the sides of the body. In the common earth-worm there are no less than one hundred and twenty of these minute air vesicles, each of which is provided with an external opening placed between the segments of the body. In the leech, the number is reduced to sixteen on each side, which open externally by the same number of minute orifices. Over the internal surface of these air vesicles the blood of the system is distributed in minute and delicate capillaries; and is capable of being aërated by whichever medium may pass through the external orifices, whether water or air.

351. In this simple apparatus is apparent the rudiment of the more perfect aërial respiration by the organs termed tracheæ, minute air tubes which ramify like blood-vessels through the body (fig. cxxx. 3). These air tubes open on the external surface by distinct apertures termed *spiracula* or *spiracles* (fig. cxxx. 2), which are commonly placed in rows on each side of the body (fig. cxxx. 2), with distinct prominent edges (fig. cxxx. 2), often surrounded with hairs; sometimes guarded by valves to prevent the entrance of extraneous bodies, and capable of being opened and closed by muscles specially provided for that purpose. These tubes, as they proceed from the spiracles to be distributed to the different organs of the body, often present the appearance of radii (fig. cxxx. 3), and when traced to their terminations are found

to end in vesicles of various sizes and figures, but commonly of an elongated and oblong form. These minute vesicles, when examined by the microscope, are seen to afford still minuter ramifications, which are ultimately lost in the tissues of the body.

352. The tracheæ are composed of three tunics, the external dense, white and shining; the internal soft and mucous, between which is placed a middle tunic, dense, firm, elastic, and coiled into a spiral. By this arrangement the tube is constantly kept in a state of expansion, and is therefore always open to the access of air. A great part of the blood of the body, in the extensive class of creatures provided with this form of respiratory apparatus, including the almost countless tribes of insects, is not contained in distinct vessels, but is diffused by transudation through the several organs and tissues of the body. All the creatures of this class live in air, and possess great activity; they therefore require a high degree of respiration; yet they are commonly small in size, and often some portions of their body consist of exceedingly dense and firm textures; hence to have localized the function of respiration, by placing the seat of it in a single organ, would have been impossible, on account of the disproportionate magnitude which such an organ must have possessed; in this case it was easier to carry the air to the blood, than the blood to the air; and accordingly the air is carried to the blood, and,

Fig. CXXXI.—*Respiratory Organs of the Scorpion.*

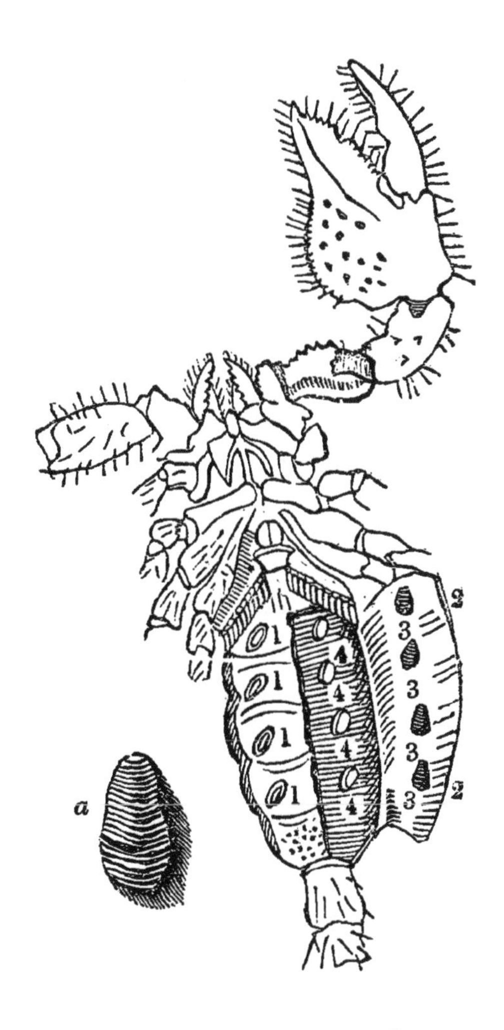

1. Spiracles. 2. Integument of one half of the body turned back. 3. Branchial organs. 4. Cells or pouches in which they are lodged. *a.* One of the respiratory organs removed and magnified, showing its resemblance to the branchial leaflets, and presenting the pectinated appearance described in the text.

like the blood in creatures of higher organization, is diffused through every part of the system.

353. The next advancement in the ascending scale is, by a step which obviously connects this higher class with the classes below and above it. It consists of distinct cells, termed pulmonic cavities (fig. CXXXI. 4), which communicate externally by spiracula (fig. CXXXI. 1), like tracheæ (351), but which are lined internally by a soft and delicate membrane plaited into folds, disposed like the teeth of a comb (pectinated) (fig. CXXXI. *a*), presenting a striking analogy to the structure of gills (345), and therefore called by the French writers pneumo-branchiæ. These cavities have the internal form of an aquatic organ, but they perform the function of air-breathing sacs. In scorpions (fig. CXXXI. 1) and spiders, this form of the apparatus is seen in its simplest condition; in the slug and snail it is more highly developed: for in these latter animals a rounded aperture, placed

Fig. CXXXII.—*Apparatus of Respiration in the Frog.*

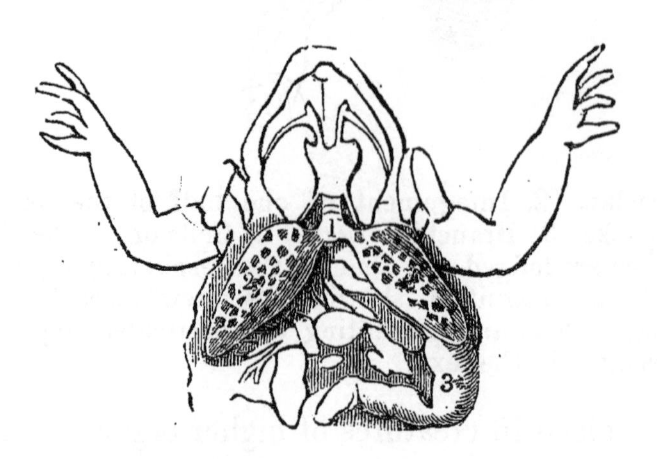

1. Trachea. 2. Vesicular lungs. 3. Stomach.

near the head, and guarded by a sphincter muscle, that alternately dilates and contracts, leads to a single cavity, which is lined with a membrane delicately folded, and overspread with a beautiful net-work of pulmonary blood-vessels.

354. Passing from this to the lowest order of the air-breathing vertebrata (fig. CXXXII.), the apparatus is perfectly analogous, but more developed. In the reptile, this air-breathing sac, which now constitutes a true and proper lung, instead of being simple and undivided, is formed by numerous septa, which traverse each other in all directions, into vesicles or cells (fig. CXXXII. 2), which proportionally enlarge the surface for the distribution of blood-vessels. In the Batrachian reptile, as the frog, salamander, newt, &c. (fig. CXXXII.), the vesicles, comparatively few in number, are of large size, and as thin and delicate as soap-bubbles. In the ophidian reptile, as the serpent, the sac is large and elongated, but divided only in the upper and back part into vesicles; while in the Saurian reptiles, as the crocodile, lizard, chamelion, &c., the sac is comparatively small, but subdivided into very minute vesicles, bearing a close analogy to the more perfectly organized lung of the higher animals.

355. In birds, the next order of vertebrata (fig. CXXXIII.), as in insects, the class of invertebrated animals which are formed for flight (352), the respiratory organs extend through the greater part

Fig. CXXXIII.—*Respiratory Apparatus of the Bird, as seen in the Swan.*

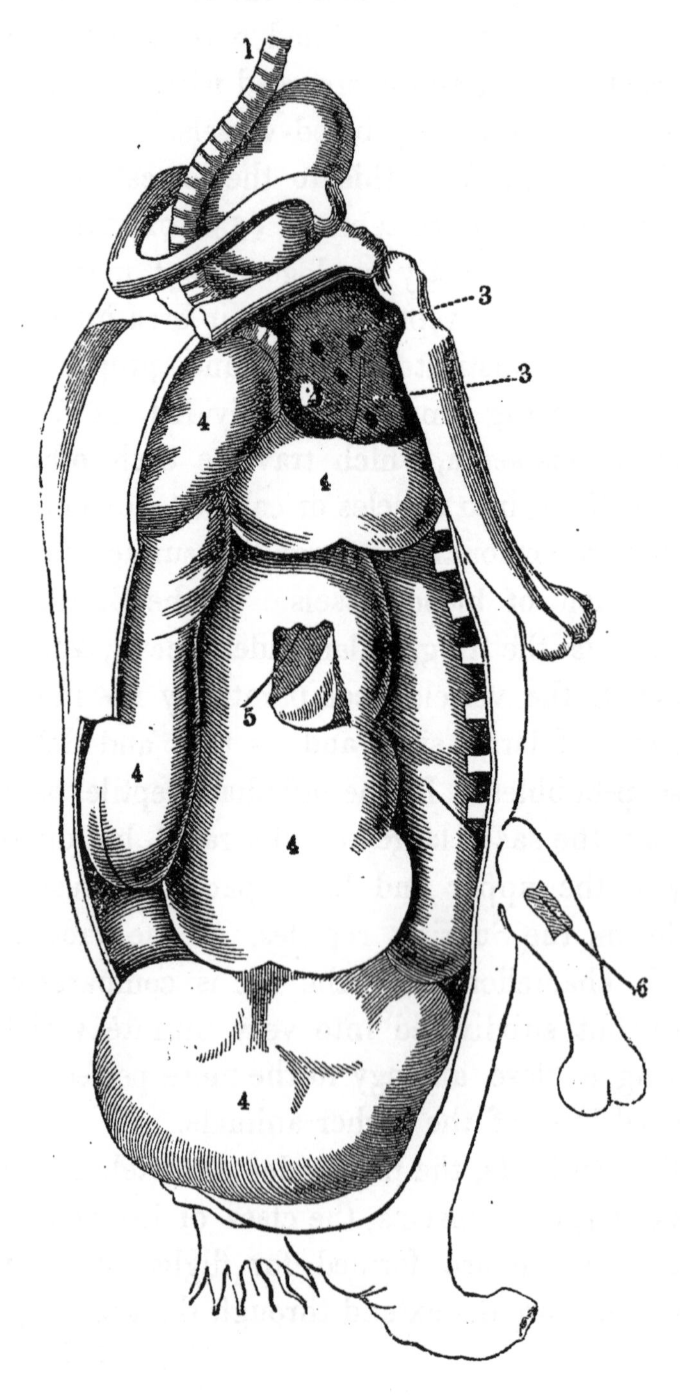

1. The Trachea. 2. The lungs. 3. Apertures through which air passes into, 4. Air cells of the body. 5. A bristle passed from one of the air cells of the body, to the cavity containing the lungs. 6. A bristle passed from the cavity of the thigh-bone into another air cell of the body.

of the body (fig. CXXXIII. 4). The lungs (fig. CXXXIII. 2), which still consist of a single pulmonic sac on each side (fig. CXXXIII. 2), are divided into cells, minute compared with those of the reptile, yet large compared with those of the quadruped; at the same time numerous air sacs, similar in structure to those of the lungs, but of larger size, are distributed over different parts of the body (fig. CXXXIII. 4), which communicate with the air cells of the lungs (fig. CXXXIII. 3); while of these larger sacs, several communicate also with the bones (fig. CXXXIII. 6), so as to fill with air those cavities which in other animals are occupied with marrow.

356. In the mammalia, the highest order of the vertebrata, respiration is less extended through the system, and is concentrated in a single organ, the lung, which, though comparatively smaller in bulk than in some of the lower classes, is far more developed in structure. The lung in this class consists of a membranous bag, divided into an immense number of distinct vesicles or cells, in the closest possible proximity with each other, yet not communicating, and presenting, from their minuteness, a vast extent of internal surface. This

bag is confined to a distinct cavity of the trunk, the thorax (fig. CXXXIV.), completely separated from the abdomen by the muscular partition, the diaphragm (fig. CXXXIV. 10). This organ no

Fig. CXXXIV. — *View of the Respiratory Apparatus in Man.*

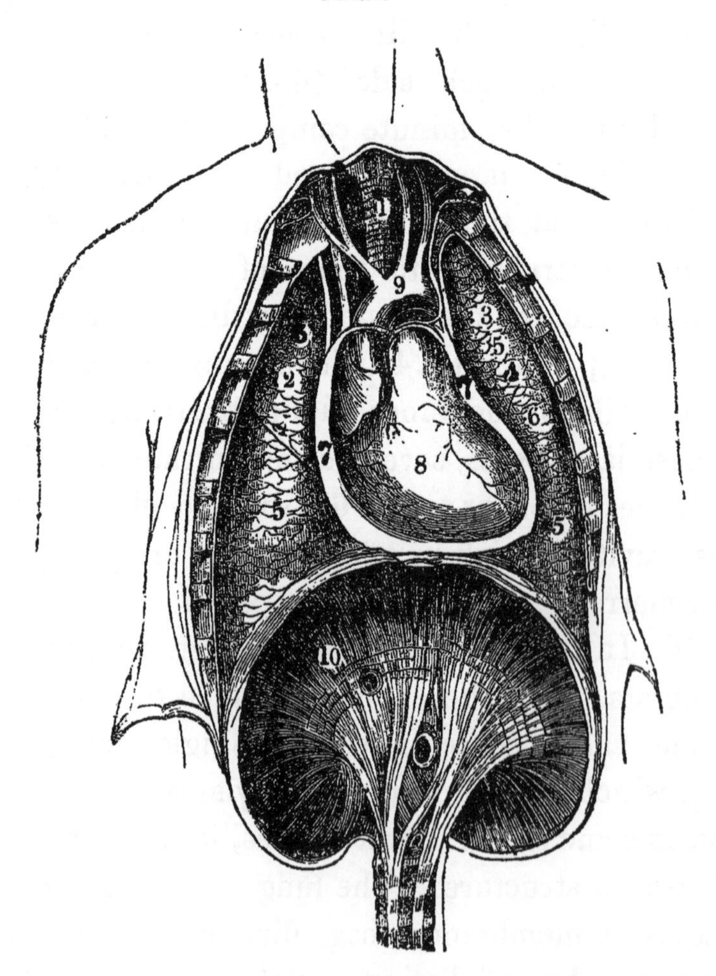

1. The Trachea. 2. The right lung. 3. The left lung. 4. Fissures, dividing each lung into, 5. Large portions termed lobes. 6. Smaller divisions termed lobules. 7. Pericardium. 8. Heart. 9. Aorta. 10. Diaphragm separating the cavity of the thorax from that of the abdomen.

longer sends down cells into the abdomen, nor membranous tubes into the bones; but is concentrated within the thorax along with the heart (fig. CXXXIV. 2, 3, 8). In all the orders of this class, the development and concentration of the organ are in strict proportion to the perfection of the general organization.

357. In man there are two pulmonary bags (fig. CXXXIV. 2, 3), of nearly equal size, which, together with the heart, completely fill the large cavity of the thorax (fig. CXXXIV.), their external surface being everywhere in immediate contact with the thoracic walls. One of these bags is placed on the right side of the body, constituting the right lung (fig. CXXXIV. 2), and the other on the left, constituting the left lung (fig. CXXXIV. 3). Each lung is divided by deep fissures, into large portions called lobes (figs. CXXXIV. 4, and CXXXV. 6), of which there are three belonging to the right, and two to the left lung. Each lobe is subdivided into innumerable smaller parts termed lobules (figs. CXXXIV. 6, and CXXXV. 6), while the lobules successively diminish in size until they terminate in minute vesicles that constitute the great bulk of the organ (fig. CXXXV. 8).

358. The complete centralization of the respiratory function which thus takes place in man, renders the apparatus exceedingly complex both on account of the expedients which are necessary to obtain the requisite extent of surface, in the

small allotted space, and to bring into contact within that space the fluids that are to act on each other.

359. The apparatus consists of a vessel to carry the air to the blood; a vessel to carry the blood to the air; an organ in which the air and the blood meet; and an organization by which both fluids are put in motion. The vessel that carries the air to the blood is the windpipe (fig. cxxxv.

Fig. CXXXV.—*View of the Air Tubes and Lung.*

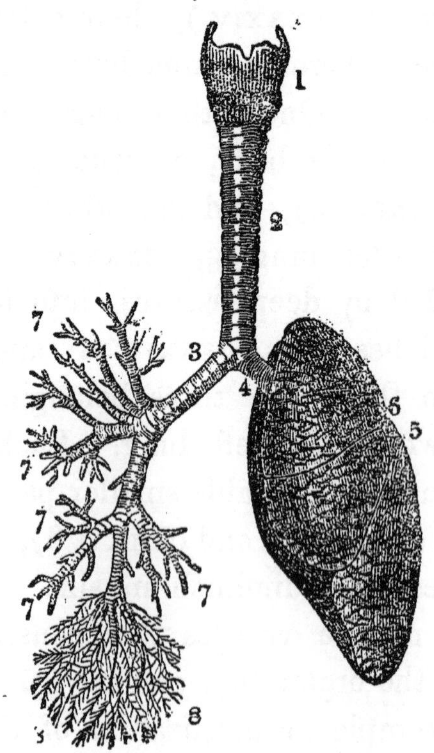

1. The larynx. 2. Trachea. 3. Right bronchus. 4. Left bronchus. 5. Left lung; the fissures denoted by the two lines which meet at 6, dividing it into three lobes, and the smaller lines on its surface marking the division of the lobes into lobules. 7. Large bronchial tubes. 8. Minute bronchial tubes terminating in the air cells or vesicles.

1, 2); the vessel that carries the blood to the air is the pulmonary artery (fig. CXL. 7); the organ in which the blood and the air meet is the lung (fig. CXXXV. 5); the organization which puts the air in motion, is the structure of bones, cartilage and muscles, called the thorax (figs. CXLI. and CXLVI.), and the engine that communicates motion to the blood is the right ventricle of the heart (fig. CXL. 5).

360. The windpipe is a tube which extends from the mouth and nostrils to the lung (figs. CLIII. 1, 9, and CXXXV. 2, 5). It is attached to the back part of the tongue (fig. CLII. 2, 9), and passes down the neck immediately before the esophagus, or the tube which leads to the stomach (fig. CLIII. 9, 12).

361. In the different parts of its course the windpipe is differently constructed, performs different offices, and receives different names according to the diversity of its structure and function. The first division of it is called the larynx (fig. CXXXV. I.), the second the trachea (fig. CXXXV. 2), the third the bronchi (figs. CXXXV. 3, 4, 7, and CXXXVII.), and the fourth the air vesicles or cells (figs. CXXXV. 8, and CXXXVIII. 2).

362. The first portion of the windpipe called the larynx (figs. CXXXV. and CXXXVI.), constitutes the organ of the voice. It is situated at the upper and fore part of the neck (fig. CLIII. 7, 9), immediately under the bone to which the root of the

tongue, called the os hyoides (figs. CLIII. 6, and CXXXVI. 1), is attached. The larynx forms a very complex structure, and is composed of a

Fig. CXXXVI.—*Posterior View of the Larynx and Trachea.*

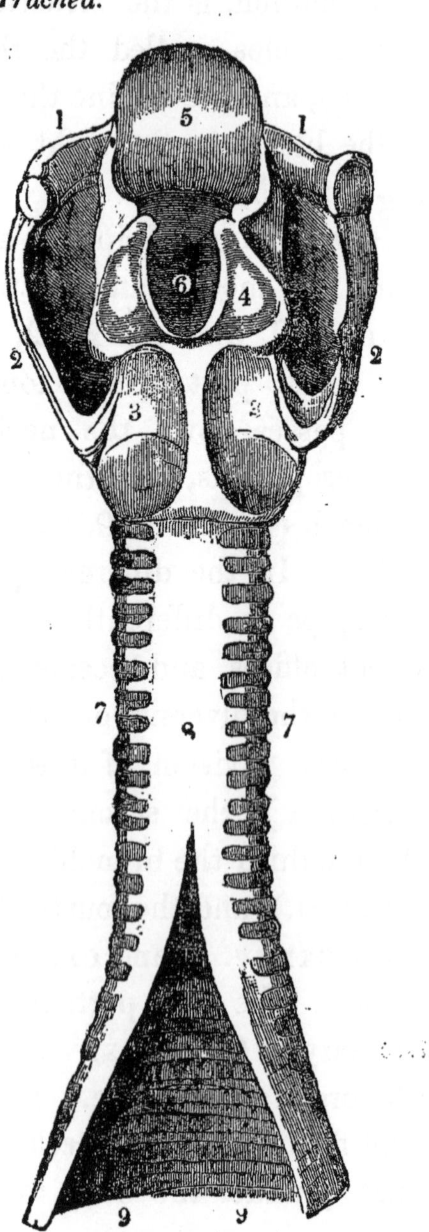

1. The os hyoides 2. Thyroid cartilage. 3. Cricoid cartilage. 4. Arytenoid cartilages, separated from each other. 5. Epiglottis. 6. Opening of the glottis. 7. Termination of the cartilaginous rings of the trachea. 8. The ligamentous portion of the trachea. 9. Trachea laid open, showing its internal mucous surface and follicles, with the anterior portion of the cartilaginous rings appearing through it.

variety of cartilages, muscles, ligaments, membranes, and mucous glands (fig. CXXXVI. 2, 3, 4, 5). At its upper part is a narrow opening of a triangular figure called the glottis (fig. CXXXVI. 6), by which air is admitted to and from the lung. Immediately above this opening is placed the cartilage, which obtains its name from its situation, *epiglottis* (fig. CXXXVI. 5), which is attached to the root of the tongue (fig. CLIII. 6, 7), and which may be distinctly seen in the living body by pressing down the tongue.

363. The Epiglottis is highly elastic, and is an agent of no inconsiderable importance in respiration, deglutition, and speaking. In respiration it breaks the current of air which rushes to the lungs through the mouth and nostrils, and prevents it from flowing to the delicate air cells with too great a degree of force. During the action of deglutition the epiglottis is carried completely over the glottis (fig. CLIII. 6, 7, 8), partly because it is necessarily forced backwards, when the tongue passes backwards in delivering the food to the pharynx (fig. CLIII. 6, 7, 8, 10), partly because it is carried backwards by certain minute muscles which act directly upon it, and perhaps also partly in consequence of its own peculiar irritability. The moment the action of deglutition has been performed, the epiglottis springs from the aperture of the glottis, partly by its own elasticity, and partly by the return of the tongue to its former position.

During the act of speaking the column of air which is expelled from the lung, which rushes through the glottis, and which thus forms the voice, strikes against the epiglottis, and the voice becomes thereby in some degree modified.

364. The second portion of the windpipe termed the trachea (fig. cxxxv. 2), commences at the under part of the larynx (fig. cxxxv. 1), and extends as far as the third dorsal vertebra, opposite

Fig. CXXXVII.

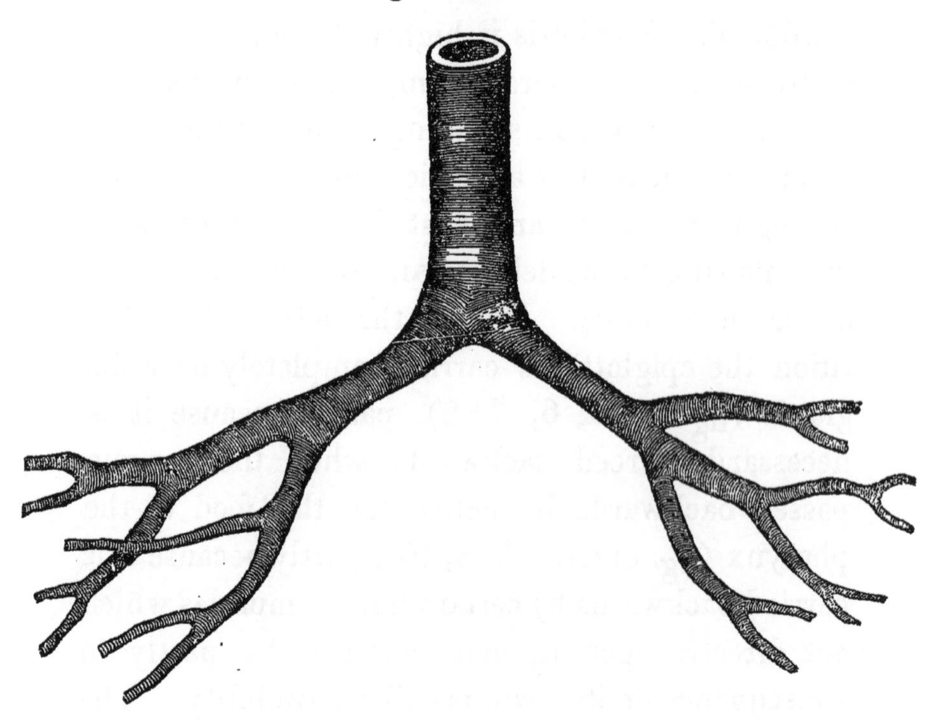

View of the trachea, showing, first, the division of the tube into the right and left bronchus, and the subdivision of the bronchi into the bronchial tubes; and secondly, the membranous and cartilaginous tissues of which the organ is composed.

to which it divides into two branches which are termed the bronchi (fig. cxxxv. 3, 4, and cxxxvii.). One of these branches, called the right bronchus, goes to the right lung; the other branch, called the left bronchus, goes to the left lung (fig. cxxxv. 3, 4).

365. The trachea of man, like the tracheæ of the air-breathing insect (351), is composed of three tissues. These tissues differ essentially from each other in nature, and are widely different in form and arrangement. They consist of membrane, muscle, and cartilage.

366. The membranous portion of the human trachea consists of three coats, the cellular (fig. cxxxvii.), the ligamentous (fig. cxxxvi. 8), and the mucous (fig. cxxxvi. 9). From the cellular and ligamentous coats the tube receives its strength, and in some degree its elasticity; and the mucous coat constitutes the chief seat of the respiratory function. Between the ligamentous and mucous coats are placed two sets of muscular fibres; the first, the external set, passes in a circular direction around the tube; the second set, placed immediately beneath the circular, is disposed longitudinally, and collected into bundles. The office of the circular fibres is to diminish the calibre of the tube, and that of the longitudinal is to diminish its length.

367. As the tracheæ of the insect are kept constantly open for the free admission of air by their

middle membranous tunic, dense, firm, elastic, and coiled into a spiral (351), so, for the accomplishment of the same purpose, there are placed between the membranous coats of the human trachea delicate rings of the more highly organized substance, cartilage (35). These cartilaginous rings amount in the entire course of the tube to sixteen or eighteen in number (fig. CXXXV. 2); each cartilage being about a line in breadth, and the fourth of a line in thickness. They never form complete circles, but only a large segment of a circle (fig. CXXXVI. 7); the circle is incomplete behind (fig. CXXXVI. 7, 9), because there the esophagus is in direct contact with the trachea (fig. CLIII. 9, 12), and instead of dense and firm cartilage, a soft and yielding substance is placed in this situation, in order that there may be no impediment to the free dilatation of the esophagus during the passage of the food.

368. The point at which the bronchi enter the substance of the lung is called the root of the lung (fig. CXXXV. 3, 4). As soon as the bronchi begin to divide and ramify within the lung, each cartilage, instead of preserving its crescent shape, is divided into two or three separate pieces, which nevertheless are still so disposed as to keep the tube open. With the progressive diminution in the size of the bronchial branches, their cartilages become less numerous, and are placed at greater distances from each other, until at length as the

bronchi terminate in the vesicles, the cartilages wholly disappear; and with the decreasing number and size of the cartilages, the thickness of the cellular, ligamentous, and muscular coats of the bronchi also lessens, until at the points where the cartilages disappear, the muscular and mucous tunics, now reduced to a state of extreme tenuity, alone remain. The essential constituent of the air vesicles, then, is the mucous membrane; but there is reason to suppose that the muscular tunic is likewise continued over these vesicles.

369. It has been stated that the tracheæ of the insect terminate in the different tissues of its body by minute vesicles of an oblong form. The termination of the bronchi in the human lung presents a strikingly analogous appearance. Malpighi, who with extraordinary talent and success devoted his life to the investigation of the minute structures of the various organs of the human body, represents the mucous membrane of the bronchial tubes as terminating in minute vesicles of unequal size: and Reisseissen, who has more recently resumed the inquiry and examined this structure with extreme care, agrees with Malpighi in stating that the bronchial tubes at their terminal points expand into minute, delicate, membranous vesicles of a cylindrical and somewhat rounded figure (fig. cxxxviii. 2). The bronchial tubes do not divide to any great degree of minuteness (fig. cxxxviii. 1), but terminate somewhat abruptly in the vesicles

(fig. CXXXVIII. 2), which though minute are large enough to be visible to the naked eye (fig. CXXXVIII. 2). Viewed in connexion with the bronchial tubes at their terminal points, the vesicles present a clustered appearance, not unlike clusters of currants attached to their stem (fig. CXXXVIII. 2).

Fig. CXXXVIII.—*View of the Bronchial Tubes terminating in Air vesicles.*

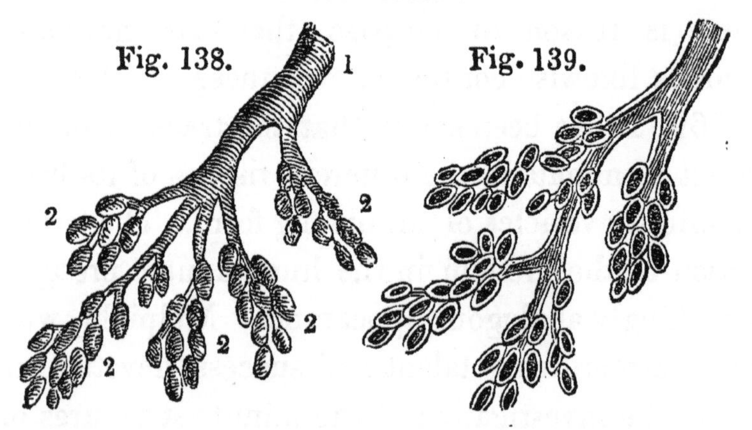

External view.—1. Bronchial tube. 2. Air vesicles. Fig. 139. The same laid open.

370. In the insect, for the reason assigned (351), these vesicles are diffused over the system, aërating every point of the body; in man they are concentrated in the lung; yet by their minuteness, and by the mode in which they are arranged, they present in the small space occupied by this organ, so extended a surface that Hales, representing the size of each vesicle at the 100dth part of an inch in diameter, estimates the amount of surface furnished by them collectively at 20,000 square

Fig. CXL.

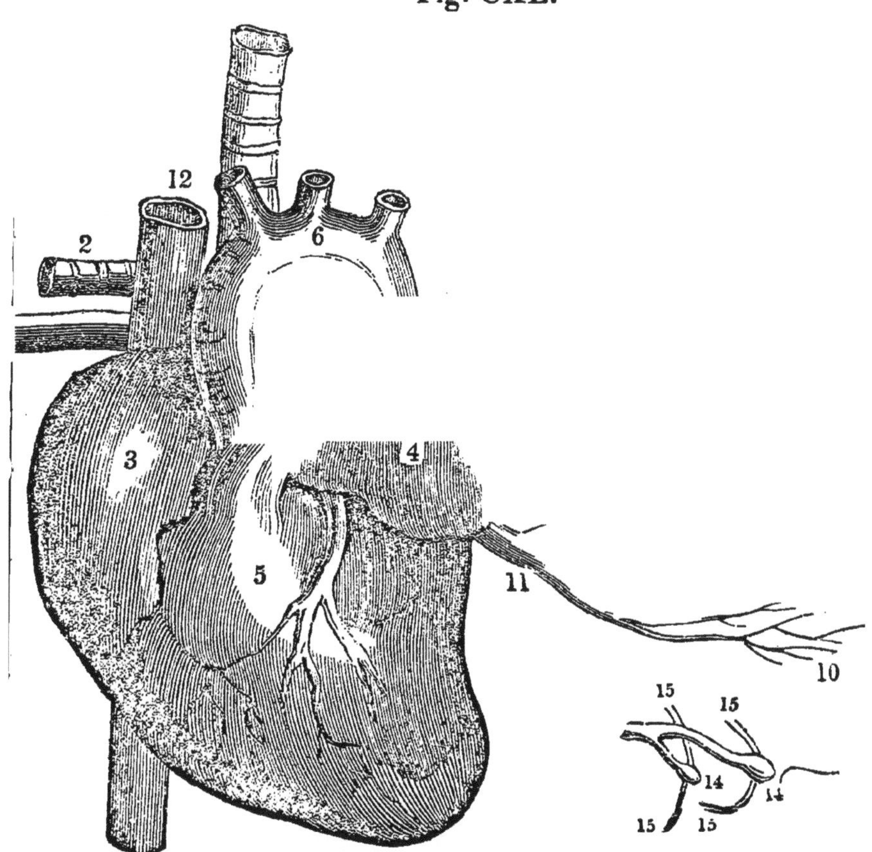

1. The trachea. 2. The right and left bronchus; the left bronchus showing its division into smaller and smaller branches in the lung, and the ultimate termination of the branches in the air vesicles. 3. Right auricle of the heart. 4. Left auricle. 5. Right ventricle. 6. The aorta arising from the left ventricle, the left ventricle being in this diagram concealed by the right. 7. Pulmonary artery arising from the right ventricle and dividing into, 8. The right, and 9. The left branch. The latter is seen dividing into smaller and smaller branches, and ultimately terminating on the air vesicles. 10. Branches of one of the pulmonary veins proceeding from the terminations of the pulmonary artery on the air vesicles, where together they form the net-work of vessels termed the Rete Mirabile. 11. Trunk of the vein on its way to the left auricle of the heart. 12. Superior vena cava. 13. Inferior vena cava. 14. Air vesicles magnified. 15. Blood-vessels distributed upon them.

inches. Keil estimating the number of the vesicles at 174,000,000, calculates the surface they present, at 21,906 square inches. Leiberkuhn at 150 cubic feet; and, according to Monro, it is thirty times the surface of the human body.

371. Such is the structure of the vessel that carries the air to the blood, and such is the mode of its distribution.

The vessel that conveys the blood to the air is the pulmonary artery, the great vessel which springs from the right ventricle of the heart (fig. CXL. 5).

The pulmonary artery soon after it issues from the right ventricle of the heart divides into two branches (fig. CXL. 7, 8, 9), one for each lung (fig. CXL. 8, 9). Each branch of the pulmonary artery as soon as it enters its corresponding lung (fig. CXL. 9) divides and ramifies through the organ in a manner precisely similar to the bronchial tubes. Every branch of the artery is in contact with a corresponding branch of the bronchus (fig. CXL. 2), divides as it divides, and accurately tracks its course throughout (fig. CXL. 2), until the ultimate divisions of the artery at length reach the ultimate vesicles of the bronchus (fig. CXL. 2, 10), upon the delicate walls of which the capillary arteries rest, expand, and ramify, forming a net-work of vessels, so complex that the anatomist who first observed it, named it the *Rete Mirabile*, the wonderful net-work, and it is still called the *Rete Mirabile Malpighi*, or the *Rete Vasculosum Malpighi* (fig. CXL. 2, 9, 10).

372. The blood which has finished its circulation through the system, returned by the great systemic veins (fig. CXL. 12, 13), to the right side of the heart (fig. CXL. 3), is driven by the right ventricle (fig. CXL. 5), into the pulmonary artery (fig. CXL. 7); by the branches of which (fig. CXL. 8, 9) it is distributed to the air vesicles of the lungs: consequently the right heart of man bears precisely the same relation to the lungs, that the single heart of the fish bears to the branchiæ; the former is a pulmonic, as the latter is a branchial heart; one half of the double heart of the more highly organized creature is employed to circulate the venous blood of the system through the lungs, as the whole of the single heart of the less highly organized animal, is employed to propel the blood through the branchiæ (368). From the capillary branches of the pulmonary artery in the Rete Mirabile (fig. CXL. 9), arises another set of vessels termed the pulmonary veins (fig. CXL. 10), which receive the blood from the venous vessels spread out on the air vesicles; for the pulmonary artery is functionally a vein, since it contains venous blood, though it is nominally an artery because it carries blood from the heart (269); and in like manner the pulmonary veins are func-

tionally arteries since they contain arterial blood, though they are nominally veins because they carry blood to the heart (272). The branches of the pulmonary arteries are larger in size and greater in number than those of the pulmonary veins, the reverse of what is observed in any other part of the body; because the pulmonary artery contains the blood which is to be acted upon by the air, while the pulmonary veins merely receive the blood which has been acted upon by the air, and the former ramifies more minutely than the latter, in order that the air may act on a larger surface of blood.

373. In the Rete Mirabile the junction of the air-vessel with the blood-vessel is accomplished. The combination of these two sets of vessels constitutes the lung; for the lung is composed of air-vessels and blood-vessels united, and sustained by cellular tissue, and inclosed in the thin but firm membrane called the pleura (104 and 105).

374. Such is the arrangement of that part of the respiratory apparatus which contains the fluids that are to act on each other. The object of the remaining portion of it is to produce the movements which are necessary to bring the fluids into contact. This is accomplished by the mechanism and action of the thorax and diaphragm (figs. CXLI. and CXXXIV. 10).

375. These organs, which invariably act in con-

cert, are so constructed and disposed, that when in action they give to the chest two alternate motions, one that by which its capacity is enlarged; and the other that by which it is diminished. These alternate movements are called the motions of respiration. The motion by which the capacity of the chest is enlarged is termed the action of inspiration; and that by which it is diminished the action of expiration.

376. The action of inspiration, or that by which the capacity of the chest is enlarged, is effected by the combined movements of the thorax and diaphragm; by the ascent of the thorax and by the descent of the diaphragm.

377. The osseous portion of the thorax, which has been fully described (69 *et seq.*), consists of the spinal column (fig. CXLI. 1), the ribs with their cartilages (fig. CXLI. 2, 3), and the sternum (fig. CXLI. 4). The soft portion of the thorax consists of muscles and membrane (figs. CXLII., CXLVI., and CXLVII.), together with the common integuments of the body. The chief boundaries of the cavity of the thorax before, behind, and at the sides, are osseous, being formed before by the sternum and the cartilages of the ribs (fig. CXLI. 4, 3); behind by the spinal column and the necks of the ribs (fig. CXLI. 1, 2); and at the sides by the bodies of the ribs. Below the boundary is muscular, being formed by the diaphragm (fig. CXLIII. 3).

378. Externally the thorax is convex and enveloped by muscle and skin; internally it is concave (fig. CXLIII. 1), and lined by a continuation of the same membrane which envelops the lungs, the pleura (104). But that portion of the pleura which lines the internal wall of the thorax is called the costal pleura (pleura costalis), in contradistinction to that which envelopes the lungs, which is termed the pulmonary pleura, or pleura pulmonalis (104). By the costal pleura, a thin but firm and strong membrane, smooth, polished, and like all the membranes of its class (serous membrane 30, *et seq.*), kept in a state of perpetual moisture and suppleness, by a fluid secreted at its surface, the movements of the thorax are facilitated, at the same time that they are prevented from injuring the delicate organs contained in it.

379. The moveable parts of the osseous portion of the thorax are the ribs and sternum. The ribs, though by one extremity tied with exceeding firmness to the spinal column by ligaments specially constructed, and admirably adapted for that purpose (figs. LVI. 1, and LVII. 1), and though attached at their other extremity by their cartilages to the sternum (fig. LVIII.), are capable of three motions, an upward, an outward, and a downward motion.

380. The ribs form a series of moveable arches, the convexity of the arches being outwards, and

Fig. CXLI. -- *View of the osseous portion of the Thorax.*

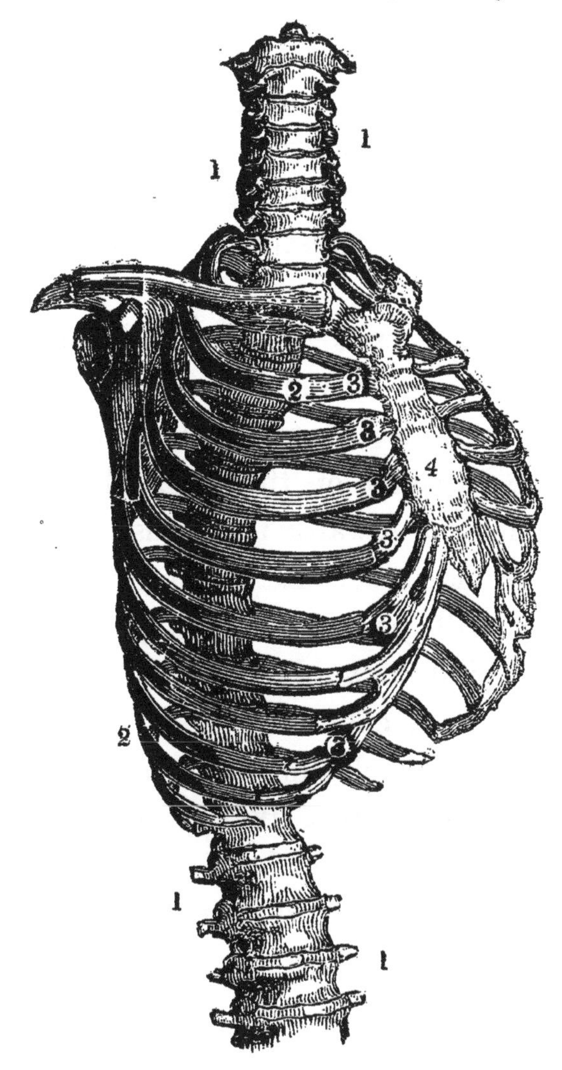

1. Spinal column. 2. Ribs. 3. Cartilages of ribs. 4. Sternum.

the whole being disposed in an oblique direction (fig. CXLI. 2). The first rib springs from the vertebral column at nearly a right angle (fig. CXLI. 2); the acuteness of this angle increases in succession as the ribs descend from the first to the last (fig. CXLI. 2); in this manner each rib is inclined obliquely outwards and downwards, and the obliquity thus given to the general direction of the ribs augments progressively from above downwards (fig. CXLI. 2).

381. In consequence of this conformation and arrangement of the ribs, every degree of motion which is communicated to them, necessarily influences the capacity of the space they enclose: If they are moved upwards they must enlarge that space at the sides, because the intervals between each other will be increased (fig. CXLI. 2); and from behind forwards, because the distance between the spinal column and the sternum (the sternum being protruded forwards with their cartilaginous extremities) (fig. CXLI. 3, 4), will be increased. If, on the other hand, they are moved downwards, the capacity of the thorax will be proportionally diminished in every direction (fig. CXLI.).

382. One part of the action of inspiration consists, then, of this ascent of the ribs. The ascent of the ribs is effected by the contraction of a double layer of muscles called the intercostal (fig. CXLII.), placed in succession between each rib; and which communicate this motion in the fol-

lowing mode. The first rib is fixed; the second rib is moveable, but less moveable than the third, the third than the fourth, and so on through the series: consequently the contraction of the intercostal muscles (figs. CXLII. and CXLVI. 2) must

Fig. CXLII.

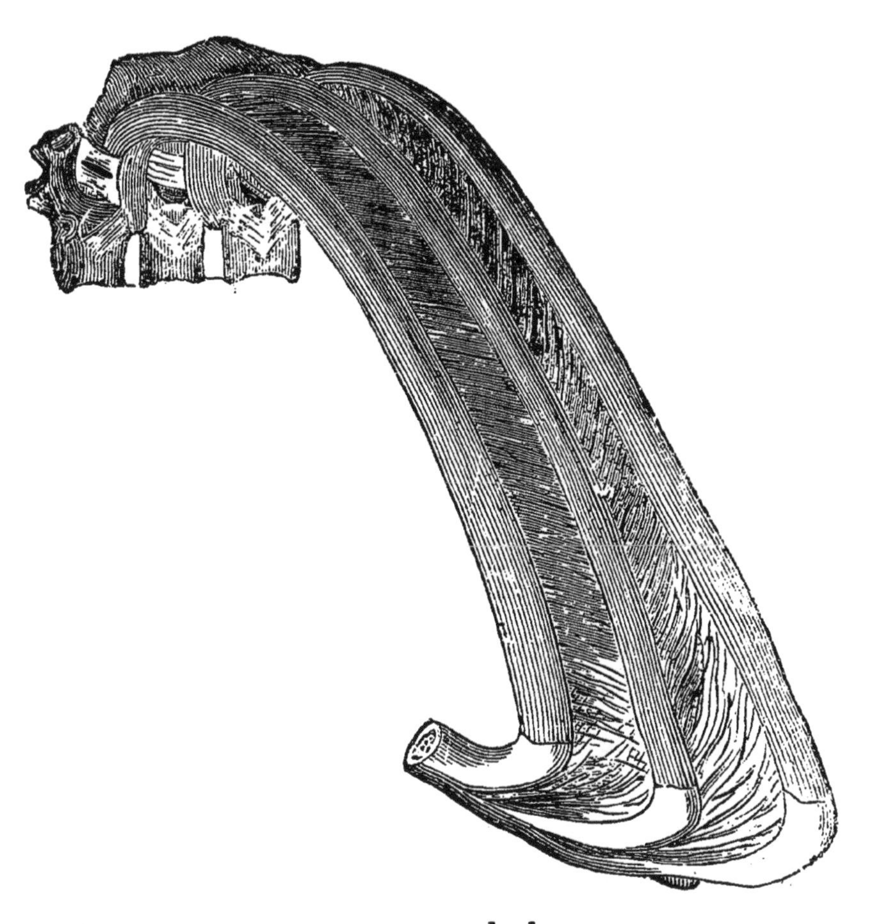

View of the intercostal muscles which fill up the interspaces between the ribs. These muscles consist of a double layer of fibres, the external and the internal, which cross or intersect each other.

elevate the whole series, because the upper ribs afford fixed points for the action of the muscles; and so, when all these muscles contract together, they necessarily pull the more moveable arches upwards towards the more fixed (figs. CXLI. and CXLVI. 2).

383. But from the oblique direction of the ribs, they cannot ascend without at the same time protruding forwards their anterior extremities (fig. CXLI.). Those extremities being attached to the sternum, which forms the anterior wall of the thorax, they cannot be protruded forwards without at the same time carrying the sternum forwards with them (fig. CXLI.). Thus, by this two-fold motion of the ribs, an upward and consequently an outward motion, the capacity of the thorax is increased from behind forwards, that is, in its small diameter.

384. Such is the part of the action, in inspiration, performed by the motion of the ribs. The remaining part of that action, by far the most important, consists of the enlargement of the capacity of the thorax from above downwards, or in its long diameter. This is effected by the descent of the diaphragm (fig. CXLIII.).

385. The diaphragm is a circular muscle, forming a complete but moveable partition between the thorax and the abdomen (figs. CXXXIV. 10, and CXLIII. 3). When not in action its upper surface

forms an arch (figs. CXLIII. 4, and CXLV. 1), the convexity of which is towards the thorax (figs. CXLIII. 4, and CXLV. 1), and reaches as high as the fourth rib (fig. CXLV. 1); its under surface, or that towards the abdomen, is concave (figs. CXXXIV. 10, and CXLV. 1). Its central portion is tendinous (fig. CXLIII. 4). This central tendinous portion of the diaphragm, which is in apposition with the heart (fig. CXXXIV. 8), and firmly attached to the pericardium (fig. CXXXIV. 7), is nearly if not quite immoveable: it is only the lateral or muscular portions (fig. CXLIII. 4) that are capable of motion. Its central portion is constructed of dense and firm tendon, and is immoveable, primarily, in order to afford one of the two fixed points (the ribs affording the other fixed point), essential to the action of the muscular fibres that constitute its lateral or moveable portions; and secondarily, in order to afford a support to the heart, which rests upon this central tendon. Thus, in consequence of this tendon being rendered absolutely fixed, the motions of the diaphragm are completely prevented from incommoding the motions of the heart; the function of respiration from interfering with the function of the circulation.

386. During the action of inspiration the muscular or lateral portions of the diaphragm contract (fig. CXLIII. 3); its muscular fibres shorten

Fig. CXLIII.—*View of the Diaphragm.*

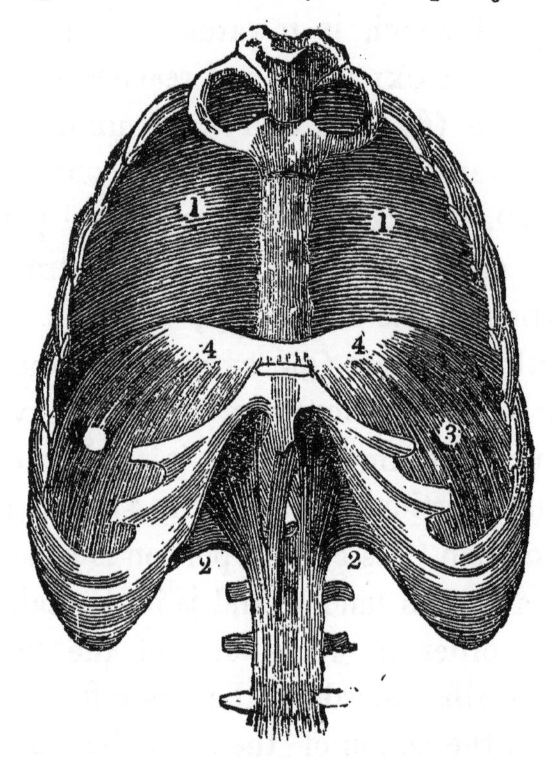

1. Cavities of the thorax. 2. Portion of cavity of the abdomen. 3. Lateral or muscular and moveable portions of the diaphragm. 4. Central or tendinous and fixed portion of the diaphragm.

themselves, and are approximated towards the central tendon (fig. CXLIII. 2); the consequence is that the whole muscle descends (fig. CXLIV. 1); passes from the fourth to below the seventh rib (fig. CXLIV.), loses its arched form and presents the appearance of an oblique plane (fig. CXLIV.). At the same time the muscles of the abdomen are protruded forwards (fig. CXLIV. 2),

Views of the Diaphragm in the different states of Respiration.

Fig. CXLIV Fig. CXLV.

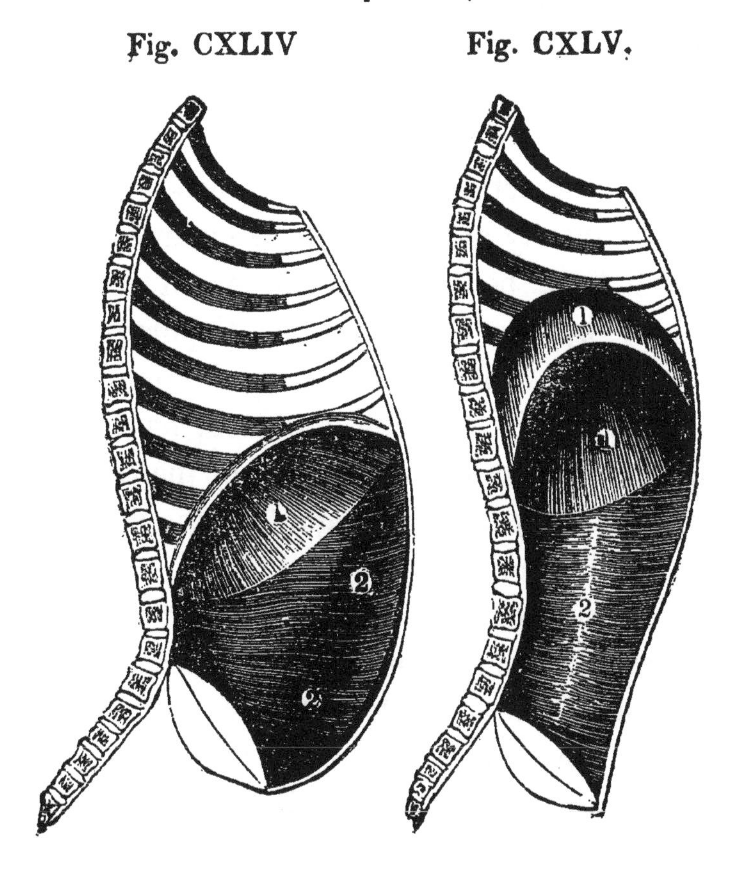

Fig. 144—1. Diaphragm in its state of greatest descent in inspiration. 2. Muscles of the abdomen, showing the extent of their protrusion in the action of inspiration. Fig. 145.—1. Diaphragm in the state of its greatest ascent in expiration. 2. Muscles of the abdomen in action forcing the viscera and diaphragm upwards.

and the viscera contained in its cavity are pushed downwards. The result of these movements is, that the capacity of the thorax is enlarged by all the space that intervenes between the fourth rib (fig. CXLV. 1), and the lowest point of the oblique plane formed by the diaphragm (fig. CXLIV. 1), together with all that gained by the protrusion of the walls of the abdomen and the descent of its viscera (fig. CXLIV. 2).

387. By the action of the intercostal muscles, then, the capacity of the thorax is enlarged at the sides and from behind forward, or in its short diameter; by the action of the diaphragm, the capacity of the thorax is enlarged from above downwards, or in its long diameter; by the combined action of both, the capacity of the thorax is enlarged in every direction, and thus the motion of inspiration is completed.

388. Expiration, the respiratory motion which alternates with that of inspiration, consists of the diminution of the capacity of the thorax, which is effected by the converse motions of the same organs; that is, by the descent of the ribs and the ascent of the diaphragm.

389. By the descent of the ribs, the capacity of the thorax is diminished in its short diameter, because by this motion, the oblique arches of the ribs are approximated to each other and to the spinal column, and the sternum is also approximated to the spinal column. The descent of the ribs is effected first by the elasticity of their cartilages (fig. CXLI. 2). When the intercostal muscles relax, the force which raised the ribs ceases to be

applied, and that moment the elasticity of the cartilages comes into play, and carries the ribs down wards. Secondly, by the contraction of the abdominal muscles (figs. CXLV. 2, and CXLVI. 6, 7, 8), the direct effect of which is to pull the ribs downwards (fig. CXLVI. 6, 7, 8).

390. By the ascent of the diaphragm the capacity of the thorax is diminished in its long diameter (fig. CXLV. 1). When the diaphragm ascends, it changes from the figure of an oblique plane (fig. CXLIV. 1), re-assumes its arched form (fig. CXLV. 1), and reaches as high as the fourth rib (fig. CXLV. 1). At the same time the abdominal muscles contract (fig. CXLV. 2), and are carried inwards towards the spinal column (fig. CXLV. 2). The result of these movements is, that the capacity of the thorax is diminished by all the space that intervenes between the lowest point of the oblique plane formed by the diaphragm and the fourth rib (fig. CXLV. 1), and by all the abdominal space lost by the contraction of the muscles of the abdomen (fig. CXLV. 2).

391. The first step necessary to the ascent of the diaphragm is the relaxation of its muscular fibres. As soon as these fibres are in a state of relaxation, that is, when the organ has changed from an active to a completely passive state, the powerful muscles of the abdomen (fig. CXLVI. 6, 7, 8) contract, and push the abdomina

Fig. CXLVI.—*View of the principal external Muscles of Respiration.*

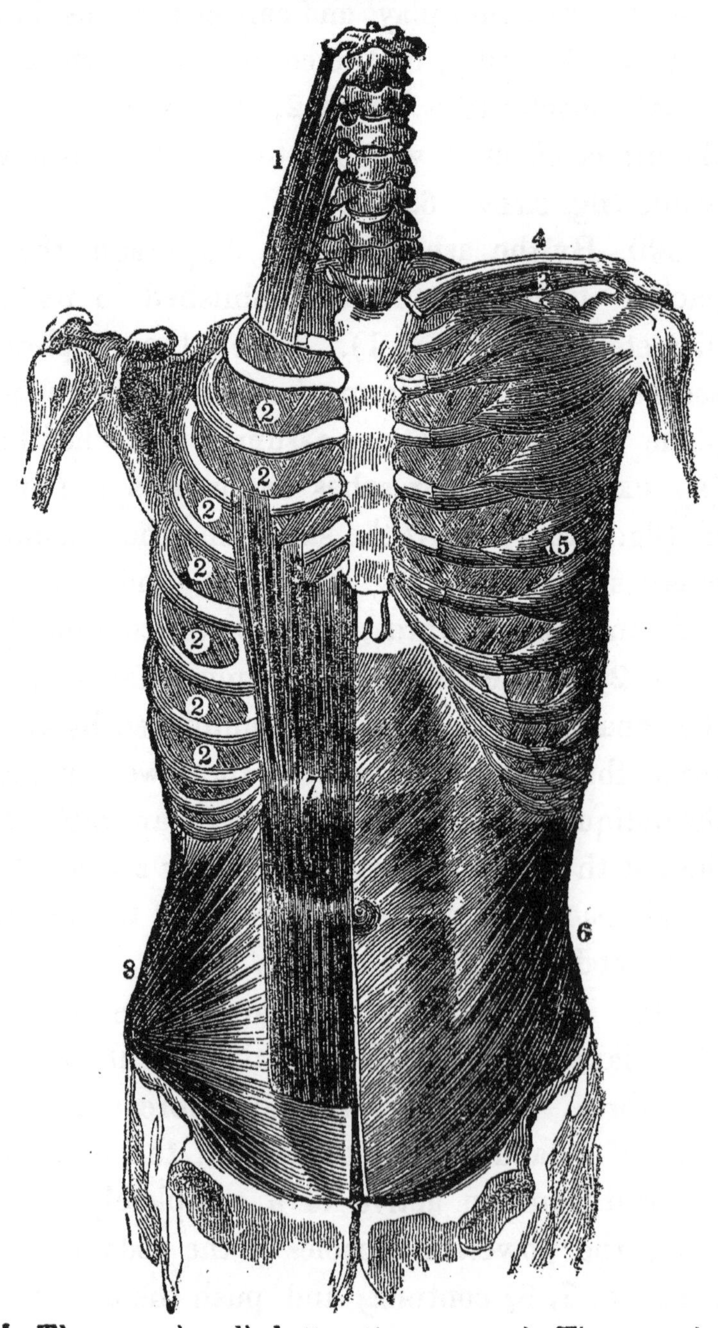

1. The muscle called the Scalenus. 2. The muscles

called the Intercostals. 3. Subclavius. 4. The bone called the Clavicle. 5. The muscle called the Serratus Magnus Anticus. 6. Obliquius Externus. 7. Rectus. 8. Obliquius Internus.

viscera and the diaphragm with them upwards towards the cavity of the chest (fig. CXLV. 2); and thus, by the descent of the ribs and the ascent of the diaphragm, the capacity of the thorax is diminished in every direction, and the motion of expiration is completed.

392. Such is the mechanism by which the capacity of the thorax is alternately enlarged and diminished in the two alternate states of inspiration and expiration, and the mechanism thus adjusted works in the following mode.

393. Expiration succeeding to the state of inspiration, the ribs descend, the diaphragm ascends, the capacity of the thorax lessens, and the compressed lungs are forced within the smallest possible space. Then, inspiration, succeeding to the state of expiration, the ribs ascend and the diaphragm descends; the capacity of the thorax is enlarged, and the lungs freed from their pressure expand and fill the greater space obtained. In about a second and a half after the state of inspiration has been induced, that of expiration recommences; the motion of inspiration occupying about double the time of the motion of expiration, and these alternate conditions succeed each other in a regular and uniform course, day and night,

during our sleeping and our waking hours to the end of life.

394. As long as the function is performed in a perfectly natural manner, a given number of these alternate movements takes place in a certain time, constituting what is termed the rythm of the respiratory motions. These motions perfectly regular in number and time, are likewise, in the natural state of the function, performed only with a certain degree of energy; but they are variously modified at the command of the will; in obedience to numerous sensations and emotions; in the performance of a great variety of complex actions, and in different states of disease. These modifying circumstances may cause the action of inspiration to be more full and deep, and that of expiration to be more forcible and complete than natural; or they may cause both movements to be shorter and quicker than common: hence the distinction of respiration into ordinary and extraordinary.

395. In ordinary respiration, that is, when the respiratory motions are perfectly calm and easy, the ascent and descent of the ribs are scarcely perceptible; the action is confined almost exclusively to the ascent and descent of the diaphragm. In this condition the only action of the intercostal muscles is to fix the ribs, and thus to afford one of the two fixed points which have been shown (385) to be essential to the action of the diaphragm. But in extraordinary respiration, that is, when circum-

stances happen in the economy which require that those motions should be extended, auxiliary sources can be put in requisition. There are many powerful muscles situated about the breast, shoulder and back (fig. CXLVI. and CXLVII.); which are capable of elevating the ribs and protruding the sternum to a very considerable extent (figs. CXLVI. 1, 2, 3, 5; and CXLVII. 1, 2, 3). Where, for example, the fullest inspiration which it is possible to take is required, the bones of the shoulder and shoulder-joint are firmly fixed by resting the hands upon the knees, and then every muscle which has the slightest connexion with the thorax, either before or behind, capable of raising the ribs, is added to the inspiratory apparatus (figs. CXLIV. and CXLVII.); at the same time that the abdominal muscles are relaxed to the utmost degree, in order to facilitate the ascent of the ribs and the descent of the diaphragm (figs. CXLIV. 2, and CXLVI. 6, 7, 8). If, on the contrary, the fullest possible expiration is required, the abdominal muscles contract most forcibly (fig. CXLV. 2), and every other muscle which is capable of still farther depressing the ribs and of elevating the diaphragm (fig. CXLVI. 6, 7, 8) is called into intense action. By these forcible and extraordinary efforts the thorax may be enlarged or diminished double its ordinary capacity.

396. Such are the mechanism and action of the

Fig. CXLVII.—*View of Muscles which are capable of assisting in elevating the Ribs and protruding the Sternum, in states of extraordinary respiration.*

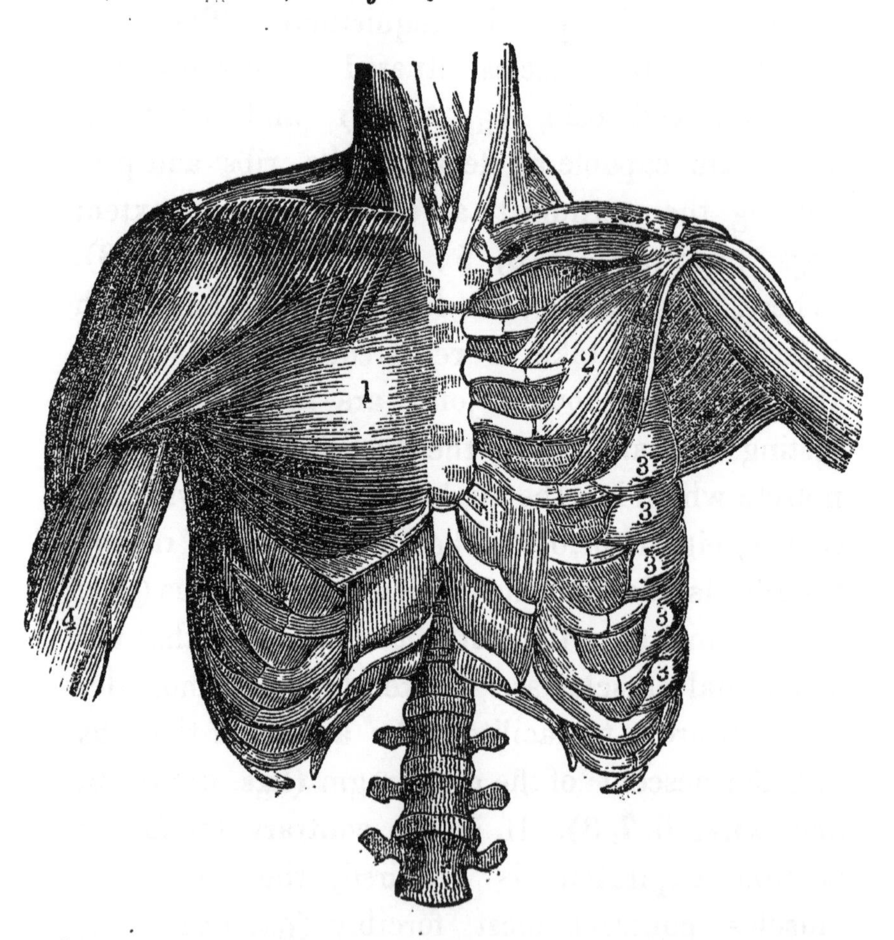

1. The muscle called the Great Pectoral. 2. The Small Pectoral. 3. The Serratus Magnus.

powers which communicate to the thorax, the motions by which its capacity is alternately enlarged and diminished, and by which the requisite impulse is communicated to the fluids which flow to and from the lungs in the different states of respiration; that is, by which air and blood flow to the lungs in the action of inspiration, and from the lungs in the action of expiration.

397. The mode in which air is transmitted to the lungs by the dilatation of the thorax, in the action of inspiration, is the following. The lungs are in direct contact with the inner surface of the thorax, and follow passively all its movements. When the volume of the lungs is reduced to its minimum by the diminished capacity of the thorax, in the state of expiration, they still contain a certain bulk of air. As their volume increases with the enlarging capacity of the thorax in the state of inspiration, this bulk of air having to occupy a greater space expands. By this expansion of the air in the interior of the lungs, it becomes rarer than the external air. Between the rarified air within the lungs, and the dense external air, there is a direct communication by the nostrils, mouth, trachea, larynx, and bronchi. In consequence of its greater weight, the dense external air rushes through these openings and tubes to the lungs and fills the air vesicles, the current continuing to flow until an equilibrium is established between the density of the air within the lungs and the density of the external air; and thus there is established the flow of a current of fresh air to the air vesicles.

398. The external air which, in obedience tc

the physical law that regulates its motion, thus rushes to the lung in order to fill the partial vacuum created by the dilatation of the thorax in inspiration, produces, in passing to the air vesicles, a peculiar sound. When the lungs are perfectly healthy, and the respiration is performed in a natural manner, if the ear be applied to any part of the chest, a slight noise can be distinguished both in the action of inspiration and that of expiration. A soft murmur, somewhat resembling the sound produced by the deep inspirations occasionally made by a person profoundly sleeping. This sound, though appreciable even by the naked ear, and though produced many times every minute, in every healthy human being from the first moment of the existence of the first man, had never been heard, or at least never attended to, until about twenty years ago, when it was observed by accident. A physician, Dr. Laennec, of Paris, having occasion to examine a young female labouring under, as he supposed, some disease of the heart, and scrupling to follow his first impulse to apply his ear to the chest, chanced to recollect that solid bodies have the power of conducting sounds better than the air. Thereupon he procured a quire of paper, rolled it up tightly, tied it, and then applied one extremity to the patient's chest and the other to his ear. Profiting by the result, which was, that he could hear the beating of the heart infinitely more distinctly than he could possibly feel it by the hand, he substituted for this first rude instrument a wooden cylinder, which he called a stethescope or chest inspector. The attentive and practised use of this instrument is found to be capable of revealing to the ear all that is passing in the chest almost as clearly and certainly as it would be visible to the eye, were the walls of the chest and the tissues of its organs transparent. Besides the entrance of the air into the lung in inspiration, and its exit in expiration, even the motion of the blood in the heart, and in the great blood-vessels, are rendered by this instrument distinctly manifest to sense; and as the ear which has once become familiar with the natural sounds produced by these operations in the state of health, can detect the slightest deviation occasioned by disease, the practical application of this discovery has already effected for the pathology of the chest, what the discovery of the circulation of the blood has accomplished for the physiology of the body.

399. At the instant that the expanding lung admits the current of air, it receives a stream of blood. The air rushes through the trachea to the air vesicles impelled by its own weight; the blood flows through the trunks of the pulmonary artery to its capillary branches, spread out on the walls of the air vesicles, driven by the contraction of the right ventricle of the heart. A current of air and a stream of blood are thus brought into so close

an approximation that nothing intervenes between the two fluids, but the fine membranes of which the air vesicles and the capillary branches of the pulmonary artery are composed, and these membranes being pervious to the air, the air comes into direct contact with the blood; the two fluids re-act on each other, and in this manner is accomplished the ultimate object of the action of inspiration.

400. On the other hand, by the action of expiration, the bulk of the lung is diminished; the air vesicles are compressed, and a portion of the air they contained, forced out of them by the collapse of the lung, is received by the bronchi, transmitted to the trachea, and ultimately conveyed out of the system by the nostrils and mouth.

401. At the same instant that a portion of air is thus expelled from the lung and carried out of the system, a stream of blood, namely, blood which has been acted upon by the air, arterial blood, is propelled from the lung and is borne by the pulmonary veins to the left side of the heart, to be transmitted to the system (fig. CXL. 10, 11, 4). In this manner, by the simultaneous expulsion from the lung of a current of air and a stream of blood is accomplished the ultimate object of the action of expiration.

402. That blood flows to the lung during the action of inspiration, and is expelled from it during the action of expiration, is established by direct experiment.

403. If the great vessel which returns the blood from the head to the heart, called the jugular vein, be exposed to view in a living animal, it is seen to be alternately filled and emptied according to the different states of inspiration and expiration.

It becomes nearly empty at the moment of inspiration, because at that moment the venous stream is hurried forward to the right chambers of the heart, which in consequence of the general dilatation of the chest are now expanded to receive it. This may be rendered still more strikingly manifest to the eye. If a glass tube, blown at the middle into a globular form, be inserted by its extremities into the jugular vein of a living animal in such a manner that the venous stream must pass through this globe, it is found that the globe becomes nearly empty during inspiration, and nearly full during expiration; empty during inspiration, because, during this action the blood flows forwards to the right chambers of the heart; full during expiration, because during this action the venous stream, retarded in its passage through the lung, its motion becomes so slow in the jugular vein that there is time for its accumulation in the glass globe. In the artery, on the contrary, in which the course of the current is the reverse of that in the vein, the opposite result takes place. In the carotid artery the stream is seen to be feeble and scanty during inspiration, but forcible and full during expiration, and if the artery be divided the jet of blood that issues from it absolutely stops during the action of inspiration; and the fuller and deeper the inspiration the longer is the interval between the jets, while it is during the action of expiration that the jet is full and strong.

404 In the course of some experiments performed by Dr. Dill and myself with a view to ascertain with greater precision the relation between respiration and circulation, we observed a phenomenon which places these points in a still more clear and striking light. We happened to divide a jugular vein. We saw that the vessel ceased to bleed during inspiration, and that it began to bleed copiously the moment expiration commenced; the reverse of what uniformly happens in the entire state of the vessel. The reason is, that the division of the vein cuts off its communication with the lung, removes it from the influence of respiration, brings it under the influence, the sole influence of the powers that move the arterial current, and consequently reverses its natural condition, and so reverses the manner in which its current flows; affording a beautiful illustration of the influence of the two actions of respiration on the two sets of blood-vessels concerned in the function.

405. It is then the venous system that is immediately related to inspiration, and the arterial to expiration. Each respiratory action exerts a specific influence over its own sanguiferous system, and the influence of the one action is the reverse

of that of the other, as the two currents they work flow in opposite directions. The lungs, in inspiration, expand and receive the venous stream; in expiration, collapse and expel the arterial stream. The expansion of the lungs in inspiration is thus simultaneous with the dilatation of the heart: during the inspiratory action both organs receive their blood. The collapse of the lungs in expiration is simultaneous with the contraction of the heart: during the expiratory action both organs expel their blood.

406. We are thus enabled to form a clear and exact conception of the mechanism and action of both parts of this complicated function. Almost all the points connected with the systemic circulation were established upwards of three hundred years ago (279), but many points connected with the pulmonic circulation have been established only recently. Our knowledge of the phenomena of both, and of their mutual relation and dependence, has been slowly increasing, and is at length tolerably complete; and now that we understand the exact office and working of each, we see that the action of the one is not only in harmony with that of the other, but co-operates with it, and renders it perfect.

407. But although the main points relative to the influence of inspiration and expiration over the pulmonary circulation may be said to be universally admitted, still physiologists are not agreed as

to the relative quantities of blood which are transmitted through the lungs during these different respiratory states. All are agreed that the state of inspiration is favourable to the passage of the blood through the lungs: some maintain that this expansion of the lungs in inspiration is essential to the pulmonary circulation. There is the like general consent that the state of expiration retards the flow of blood through the lungs; by many it is conceived that it completely stops the current. By these physiologists it is supposed that, during the action of expiration, the lungs are in a state of collapse; that they contain a comparatively small portion of air; that in this state the air vesicles are so compressed, and the pulmonary blood-vessels so coiled up, that the lungs are absolutely impermeable, and consequently, that when the blood arrives at the right chambers of the heart, it is incapable of making its way to the left. This, according to a prevalent theory, is the immediate cause of death in asphyxia, the state of the system induced by suspended respiration, as in drowning, hanging, and suffocation. Death takes place in this condition of the system, it is argued, because the circulation of the blood is arrested at the right side of the heart, cannot permeate the lungs, and consequently cannot reach the left ventricle, to be sent out to supply the organs of the body.

408. This opinion, which appears at first view

to be favoured by numerous observations and experiments, has been shown to be fallacious by a series of decisive experiments, performed by Dr. Dill and myself, undertaken, as has been stated (404), with the object of ascertaining, in a more exact manner than had hitherto been done, the relation between the circulation and respiration. The previously ascertained fact that the heart continues to beat and the blood to flow several minutes after the complete suspension of the respiration, or after apparent death, afforded us the means of pursuing our research. The details of these experiments are given elsewhere: it is sufficient to state in this place the main results.

409. As a standard of comparison, the quantity of blood which flows through the lungs after apparent death, when the lungs remain in a perfectly natural state, was previously ascertained. It was found, after death produced in an animal by a blow on the head, that blood continued to be transmitted through the lungs for the space of twenty-five minutes after the complete cessation of respiration. There passed through the lungs in all five ounces and two drachms of blood.

410. Respiration was now suspended the instant after a perfectly natural and easy *inspiration;* there flowed through the lungs four ounces and five drachms of blood.

411. Respiration was next suspended the instant after a perfectly natural and easy *expiration;*

there flowed through the lungs two ounces and seven drachms of blood.

412. When the trachea of an animal is closed by the pressure of a cord in suspension, or when an animal is immersed under water, it makes a succession of violent expirations, by which a large quantity of air is forced out of the lungs. Hence, when the lungs of an animal that has perished by hanging or drowning, are examined, they are always found much reduced in bulk; so much reduced in bulk as to have suggested the theory that the extreme collapse of the lungs and their consequent impermeability, is the cause of death in this condition of the system. On bringing this theory to the test of experiment, it was found that blood continued to flow through the lungs after apparent death from suspension, for the space of eleven minutes, and that there passed through in all five ounces of blood. The comparatively larger quantity transmitted in this case than when the inspiration and expiration were perfectly natural, was owing to the larger size of the animal. In the experiments made with a view to ascertain the relative proportions of blood transmitted through the lungs in the states of natural inspiration and expiration, the animals were chosen as nearly as possible of the same size, and were much smaller than the former.

413. On examining the quantity of blood that passed through the lungs after death from sub-

mersion, it was found to be very nearly the same as that which was transmitted after death from suspension.

414. But the lungs may be brought to a much greater degree of collapse than that to which they are reduced in hanging and drowning. By introducing an exhausting syringe into the trachea, a much larger quantity of air may be drawn out of the lungs than they are capable of expelling by the most violent efforts of expiration. When, in this mode, the lungs had been reduced to the greatest possible degree of collapse, and had been exhausted of all the air that could be drawn out of them, there flowed through them two ounces of blood.

415. Such are the results when the lungs are reduced successively from the moderate degree of collapse incident to a perfectly natural expiration, to the great degree of collapse incident to suspension and submersion, and the most extreme degree of collapse which it is possible to induce by exhaustion.

416. When the phenomena that take place in the opposite condition of the lungs were investigated, results were obtained which present a striking contrast to those which have been stated. On forcing into the lungs the largest quantity of air which they are capable of containing without the rupture of the air vesicles, and in this manner communicating to them the greatest degree of

dilatation compatible with their integrity, it was found that in this state there passed through them *only three drachms of blood.*

417. But on fully distending the lungs with water instead of air, the pulmonary circulation was instantaneously and completely arrested; they were incapable of transmitting a single drop of blood. On cutting the aorta across, as in all the preceding experiments, not a particle of blood was obtained, excepting what issued at a single jet, and which consisted only of the blood contained in the vessel at the moment the respiration was stopped.

418. From these experiments it follows—

1. That the state of inspiration is favorable to the passage of the blood through the lungs. In the dilatation of inspiration they transmitted nearly double the quantity that passed in the collapse of expiration; or, as four ounces and five drachms are to two ounces and seven drachms (410 and 411).

2. That no degree of collapse to which the lungs can be reduced is capable of wholly stopping the flow of the blood through them. In the collapse of suspension and submersion they transmitted as much blood, with the exception of two drachms, as when death was produced by a blow on the head (412 and 409). In the greatest degree of collapse capable of being produced by an

exhausting syringe, they transmitted half as much as in the collapse of suspension and submersion (414 and 412).

3. That it is only a moderate degree of dilatation that is favorable to the transmission of the blood through the lungs. When the lungs are over-distended with air, they are capable of transmitting only an exceedingly small quantity of blood (416); when they are fully distended with water, they are incapable of transmitting a single drop of blood (417). In fact they can contain only a certain quantity of air and blood; and when either of these fluids preponderates, it can only be by the proportionate exclusion of the other. It will appear hereafter that these results are capable of applications of the highest interest and importance in the explanation of numerous phenomena of health and of disease.

419. Physiologists have laboured with great diligence to determine the exact quantity of air and blood which enters and which flows from the lung at each of the actions of respiration, and they have succeeded in obtaining tolerably precise results.

420. The quantity of air capable of being received into the lungs of an adult man, in sound health, at an inspiration, is determined with correctness by an instrument constructed by Mr. Green, analagous to one suggested by Mr. Abernethy. It consists of a tin trough, about a foot

square, and six inches deep, three parts of which are filled with water. Into this trough is placed a three-gallon glass jar, open at the bottom, and graduated at the side into pints, half-pints, &c. To the upper end of the jar a flexible tube is affixed, having at its connexion a stop-cock. The lungs being emptied, as in the ordinary action of expiration, and the mouth applied to the end of the flexible tube, the nostrils being closed by the pressure of the fingers, the air is drawn out of the jar into the lungs by the ordinary action of inspiration. When as much air is thus drawn into the lungs as the air vesicles will hold, the stop-cock is closed, and the quantity of air inspired is ascertained by the rise of the water, the level of the water corresponding with the indications marked on the side of the jar.

421. The quantity of air which a person by a voluntary effort can inspire at one time is found, as might have been anticipated, to be different in every different individual. These varieties depend, among other causes, on the greater or less development of the trunk, on the presence or absence of disease in the chest, on the degree in which the lung is emptied of air by expiration previously to inspiration, and on the energy of the inspiratory effort. The greatest volume of air hitherto found to have been received by the lung, on the most powerful inspiration, is nine pints and a quarter. The average quantity which the lungs are capable

of receiving in persons in good health, and free from the accumulation of fat about the chest, appears to be from five to seven pints. The latter is about the average quantity capable of being inspired by public singers.

422. But these measurements relate to the greatest volume of air which the lungs are capable of receiving, on the most forcible inspiration which it is possible to make, after they have been emptied by forcible expiration, and consequently express the quantity received in extraordinary, not in ordinary inspiration. The quantity received at an inspiration easy, natural, and free from any great effort, may be two pints and a half; but the quantity received at an ordinary inspiration, made without any effort at all, is, according to former observations which referred to Winchester measure, about one pint.

423. The quantity of air expelled from the lung by an ordinary expiration is probably a very little less than that received by an ordinary inspiration (456).

424. No one is able by a voluntary effort to expel the whole contents of the lungs. Observation and experiment lead to the conclusion that the lungs, when moderately distended, contain at a medium about twelve pints of air. As one pint is inhaled at an ordinary inspiration, and somewhat less than the same volume is expelled at an ordinary expiration (456), there remain present

in the lungs, at a minimum, eleven pints of air. There is one act of respiration to four pulsations of the heart; and, as in the ordinary state of health there are seventy-two pulsations, so there are eighteen respirations in a minute, or 25,920 in the twenty-four hours.

425. About two ounces of blood are received by the heart at each dilatation of the auricles; about the same quantity is expelled from it at each contraction of its ventricles; consequently, as the heart dilates and contracts seventy-two times in a minute, it sends thus often to the lungs, there to be acted upon by the air, two ounces of blood. It is estimated by Haller that 10,527 grains of blood occupy the same space as 10,000 grains of water, so that if one cubic inch of water weigh 253 grains, the same bulk of blood will weigh 266⅛ grains.

426. It is ordinarily estimated that on an average one circuit of the blood is performed in 150 seconds; but it is shown (451 and 452) that the quantity of air always present in the lungs contains precisely a sufficient quantity of oxygen to oxygenate the blood, while flowing at the ordinary rate of 72 contractions of the heart per minute, for the exact space of 160 seconds. It is therefore highly probable that this interval of time, 160 seconds, is the exact period in which the blood performs one circuit, and not 150 seconds, as former observations had assigned. If this be so, then 540 circuits are performed in the

twenty-four hours; that is, there are three complete circulations of the blood through the body in every eight minutes of time.

427. But it has been shown (425) that the weight of the blood is to that of water as 1.0527 is to unity, and that consequently 10,527 grains of blood are in volume the same as 10,000 grains of water.

428. From this it results that if in the human adult two ounces of blood are propelled into the lungs at each contraction of the heart, that is, 72 times in a minute, there are in the whole body precisely 384 ounces, or 24 pounds avoirdupois, which measure 692.0657 cubic inches, or within one cubic inch of 20 imperial pints, which measure 693.1847 cubic inches.

429. By an elaborate series of calculations from these data Mr. Finlaison has deduced the following general results:—

1. As there are four pulsations to one respiration (424), there are 8 ounces of blood, measuring 14.418 cubic inches, presented to 10.5843 grains of air, measuring 34.24105 cubic inches.

2. The whole contents of the lungs is equal to a volume of very nearly 411 cubic inches full of air, weighing 127 grains, of which 29.18132 grains are oxygen.

3. In the space of five-sixth parts of one second of time, two ounces, or 960 grains weight

of blood, measuring 3⅗ or 3.60451 cubic inches, are presented for aëration.

4. Therefore the air contained in the lungs is 114 times the bulk of the blood presented, while the weight of the blood so presented is 7½ times as great as the weight of the air contained.

5. In one minute of time the fresh air inspired amounts to 616⅓ cubic inches, or as nearly as may be 18 pints, weighing 190½ grains.

6. In one hour the quantity inspired amounts to 1066⅔ pints, or 2 hogsheads, 20 gallons, and 10⅔ pints, weighing 23¾ ounces and 31 grains.

7. In one day it amounts to 57 hogsheads, 1 gallon, and 7¼ pints, weighing 571½ ounces and 25 grains (454).

8. To this volume of air there are presented for aëration in one minute of time 144 ounces of blood, in volume 259½ cubic inches, which is within 18 cubic inches of an imperial gallon.

9. In one hour 540 pounds avoirdupois, measuring 449¼ pints, or I hogshead and 1¼ pints;—and

10. In the twenty-four hours, in weight 12,960 pounds; in bulk 10,782½ pints, that is, 24 hogsheads and 4 gallons.

11. Thus, in round numbers, there flow to the human lungs every minute nearly 18 pints of air (besides the 12 pints constantly in the air vesicles) and nearly 8 pints of blood; but in the space of

twenty-four hours, upwards of 57 hogsheads of air and 24 hogsheads of blood.

430. Provision cannot have been made for bringing into contact such immense quantities of air and blood, unless important changes are to be produced in both fluids; and accordingly it is found that the air is essentially changed by its contact with the blood, and the blood by its contact with the air.

431. Chemistry has demonstrated the changes effected in the air. Common atmospheric air is a compound body, consisting of pure air and of certain substances diffused in it. Pure air is composed of two gases, azote and oxygen, always combined in fixed proportions. The substances diffused in pure air, and which are in variable quantity, are aqueous vapour and carbonic acid gas. These latter substances form no part of the chemical agents essentially concerned in the process of respiration. The only constituents of the air which are essentially concerned in the process of respiration are the two gases, azote and oxygen, the union of which, in definite proportions, constitutes pure air. But of these two gases each does not perform the same part in the function of respiration, nor is each equally necessary to the support of life.

432. If a living animal be placed in a vessel full of atmospheric air, and if all communication of the atmosphere with the vessel be prevented,

the animal in a given time perishes. If an animal be placed in a vessel full of azote, after a given time it equally perishes; but if an animal be placed in a vessel full of oxygen, not only is the function of respiration carried on with far greater energy than in atmospheric air, but the animal lives a much longer time than in the same bulk of the latter fluid. If twenty cubic inches of pure oxygen be capable of sustaining the life of an animal for the space of fourteen minutes, it can support life in the same bulk of atmospheric air only six minutes; and if its respiration be confined to either of these gases, after they have been already respired by another animal of the same species, the former will live only four minutes; that is, not longer than when entirely deprived of air. It follows that the gas which gives to atmospheric air its chief power of sustaining life is oxygen.

433. Accordingly it is proved that no animal, from the lowest to the highest, is capable of sustaining life unless a certain proportion of oxygen be present in the fluid which it respires. Whether it breathe by the skin, by gills, or by lungs, whether the respiratory medium be water or air, the presence of oxygen is alike indispensable. Yet the life of no animal can be sustained by pure oxygen. If azote be not mixed with oxygen, evils are produced in the economy which sooner or later prove fatal. On the other hand, if the proportion

of oxygen be diminished beyond a certain point, drowsiness, torpor, and death result. Not oxygen alone, then, but oxygen combined with azote, in the proportion in which nature has united these two fluids to form the atmosphere of the globe, is indispensable to animal existence.

434. When the same portion of atmospheric air is repeatedly respired by an animal, the oxygen contained in it gradually disappears, the gas lessening with every successive respiration, until at last so small a quantity remains that it is no longer capable of sustaining the life of an animal of that class. When respiration has deprived the air of its oxygen to such an extent, that it can no longer support animal life, the air is said to be consumed; but, correctly speaking, it is merely changed in composition, in the proportions in which its constituents are combined; consequently the effect of respiration is to alter the chemical composition of the air.

435. The essential change that takes place consists in the diminution of the oxygen and the increase of the carbonic acid. When inspired, atmospheric air goes to the lungs loaded with oxygen; when expired, it returns loaded with carbonic acid That the air which returns from the lungs is loaded with carbonic acid, may be rendered manifest even to the eye. If a person breathe through a tube into water holding lime in solution, the carbonic acid contained in the ex-

pired air will unite with the lime and form a white powder analogous to chalk (carbonate of lime), which being insoluble, becomes visible.

436. On the other hand, the diminution of oxygen is demonstrated by chemical analysis. If 100 parts of atmospheric air be successively respired, until it is no longer capable of supporting life, and if it be then subjected to analysis, it is found that in place of being composed of 79 parts azote, 21 oxygen, and a variable quantity of carbonic acid, sometimes amounting to half a grain per cent., it consists of 77 parts azote, and 23 carbonic acid. The oxygen is gone, and is replaced by 23 parts of carbonic acid; at least this is the ordinary estimate; but different experimentalists differ somewhat in their account of the absolute quantity of oxygen that disappears, and of carbonic acid that is generated.

437. Whatever estimates of the oxygen consumed, and of the carbonic acid generated, be adopted, they can be taken only as medium quantities. Dr. Edwards has demonstrated that the absolute quantity of oxygen consumed in a given time is constantly varying, not only in animals of different species, but even in the same animal under different circumstances; insomuch, that there are scarcely two hours in the day in which the same individual expends precisely the same quantity. The nature and degree of the exercise taken during the observation, the condition of the mind,

the state of the health, the kind of food, the temperature of the air, and innumerable other causes materially influence the quantity of oxygen consumed. When, for example, the hourly consumption of oxygen, at the temperature of 54° Fahrenheit, amounted to 1345 cubic inches,* it fell, at the temperature of 79°, to 1210 cubic inches. During the process of digestion more is consumed than when the stomach is empty; more is required when the diet is animal than when it is vegetable, and more when the body and mind are active than when at rest.

438. With regard to the carbonic acid, Dr. Prout has recently made the remarkable discovery, not only that the generation of this gas differs according to different circumstances, and more especially according to particular states of the system; but that the quantity of it which is produced regularly varies at particular periods of the day. The quantity generated is always more abundant during the day than during the night. About daybreak it begins to increase; continues to do so until noon, when it comes to its maximum, and then decreases until sunset. The maximum quantity generated at noon exceeds the minimum by about one-fifth of the whole. If from any cause the relative quantity be either increased or diminished above or below the ordinary maximum or minimum, it is invariably

* The ordinary consumption of oxygen is, for an adult, 1905 cubic inches per hour (444).

diminished or increased in an equal proportion during some subsequent diurnal period. The absolute quantity generated is materially diminished by the operation of any debilitating cause, such as low diet, protracted fasting, or long-continued exercise, depressing passions and the like. Few circumstances of any kind increase the quantity produced, and those only in a slight degree.

439. The changes produced by respiration on the other constituent of the air, azote, appear at first view to be extremely variable. By numerous and accurate experiments it is established that the quantity of this gas is at one time increased; at another diminished, and at another unchanged. It is probable that there is a constant absorption and exhalation of it; and that the apparent irregularity is the result of the preponderance of the one process over the other. When absorption preponderates, a smaller quantity is found in the air expired than in that inspired: when exhalation preponderates, a larger quantity is expired than inspired; and when the absorption and exhalation are equal, just as much is expired as inspired, and consequently there appears to be no absorption at all.

440. Such are the phenomena of respiration, as far as the labours of physiologists has succeeded in ascertaining them, up to the present time. But as the estimates of the quantity of air and blood contained in the lungs were rather matters of con-

jecture than of demonstration, and as the quantity of oxygen consumed, of carbonic acid generated, and of azote absorbed, appeared still not to be determined with exactness, I requested Mr. Finlaison to apply his power of calculation to the investigation of this subject, taking as the basis of his calculations the facts positively and precisely ascertained by experiment and analysis. This he has done with great care, and has obtained the following results.

441. It was formerly estimated that the weight of pure atmospheric air is 305,000 grains troy for one million of cubic inches; but the latest authorities assign it to be 310,117 grains. Of this weight of one million of cubic inches of pure air,

The weight of the oxygen is - - -	71,809.3
The weight of the azote is - - - -	238,307.7
Total - - - - - - - - -	310,117.0

442. But common atmospheric air in its ordinary state contains in 1000 cubic inches,

Of pure air - - - - - - - - - -	989
Of the vapour of water - - -	10
Of carbonic acid gas - - - - -	1

Ten inches of pure air are equal in weight to nine of oxygen.

Eight inches of azote are equal in weight to seven of oxygen.

The specific gravity of carbonic acid is to pure air at the rate of 15,277 to 10,000.

The specific gravity of the vapour of water is to pure air as 6,230 to 10,000. It follows that a million of cubic inches of air in its ordinary state weigh 309,111½ grains.

Carbonic acid gas is composed of oxygen and pure carbon in the proportion of eight grains of oxygen to three of carbon out of every eleven grains of carbonic acid.

443. Though during particular portions in the twenty-four hours, under circumstances which influence variously the actions of life (437 and 438), the quantity of the oxygen consumed, of carbonic acid generated, and of azote absorbed, vary (436 to 439), yet it is probable that the daily consumption, reproduction, and absorption of these gases, is pretty much the same one day with another. The experiments of Dr. Edwards clearly show that while these quantities vary to such an extent, when the observation embraces only a short interval, as to be scarcely ever the same hour by hour, yet that they lessen as the interval extends, until at length a nearly exact equilibrium is established.

444. Experimental philosophers have not obtained precisely the same results as to the quantities consumed and reproduced of these respective gases. At present, therefore, we can only approximate to the exact amount by taking the average of their observations. The following are the results of the principal experiments which

have been instituted. The quantity of oxygen consumed by an adult man in twenty-four hours is, according to

Menzes - - - - - - - -	51,840
Lavoisier - - - - - - -	46,048
Davy - - - - - - - - - -	45,504
Allen and Pepys - - -	39,534

The mean of all which is, 45,731.5 inches.

445. In like manner the quantity of carbonic acid generated in the same time is, according to

Davy - - - - -	38,304 cubic inches.
Allen and Pepys -	38,232
Thermean of which is,	38,268 „

The weight of 38,268 inches of carbonic acid gas is 18,130.1474 grains troy; and the weight of 45,731½ inches of oxygen is 15,757.9131 grains troy.

Now this weight of oxygen must have been derived from the decomposition of 221,882 cubic inches of common atmospheric air.

446. It has been shown that, in the state of health, one contraction of the heart propels to the lungs two ounces of blood; that this action of the heart is repeated 72 times in one minute; that to every four actions of the heart there is one action of respiration; that consequently there are 18 respirations in a minute, and 25,920 in the twenty-four hours.

447. From these premises it results that at

each action of the heart there is decomposed of the air inspired, 8.5603 cubic inches, that is, a quarter of a pint within one-tenth of a cubic inch, —the quarter of a pint imperial measure being 8.6648 cubic inches.

448. Previous observation had assigned one pint as the volume of air ordinarily inhaled at a single inspiration. We now see that the quantity decomposed is a quarter of a pint. It is, then, an absolute truth, that of the whole volume of air inspired, one-fourth part only is decomposed, and that three-fourths, after having been diffused through the air vesicles of the lungs, are expired without change.

449. Observation had also assigned 12 pints of air as the volume constantly present in the lungs,—that is, - - - - - 415.9108 cubic inches.
The truth seems to be, that forty-eight times the quantity decomposed is constantly present, namely, 410.8926 cubic inches.
The difference is only - - 4.0182 cubic inches,
which difference weighs less than 1¼ grains troy.

450. It is then concluded that the real contents of the lungs is a volume of 410.8926 cubic inches, which is exactly the 540th part of 221,882 cubic inches, being the whole volume decomposed in twenty-four hours. But 160 seconds is also exactly the 540th part of the number of seconds in twenty-four hours.

451. Of the whole weight of oxygen consumed in twenty-four hours - - - - 15,757.9131 grains, the 540th part, or the proportion of 160 seconds, is - - - 29.18132 „ and 410.8926 cubic inches of atmospheric air, which, as above, is the contents of the lungs, contain of oxygen the same weight - - - - - - - - - 29.18132 „

452. Then, if respiration were suddenly stopped, provision is made by the quantity of air always retained in the lungs for the oxygenation of the blood while flowing at the ordinary rate of 72 strokes per minute, for the exact space of 160 seconds, and for not one instant longer.

453. This interval of time, then, as has been stated (426), is very probably the time in which the blood performs one circuit, not 150 seconds. Then 540 circuits are performed in the twenty-four hours, or 3 circuits in every eight minutes. From this estimate has been deduced the quantity of blood contained in the whole body of the human adult (428).

454. The air inspired in twenty-four hours contains as under:—

	Bulk in cubic inches.	Weight in grains troy.	Ingredients.
Undecomposed, and to be returned unchanged.....	665,646	205,758.833,	Common air,
To be decomposed, containing in solution			
Pure atmospheric air....	219,441	15,757.913,	Oxygen,
		52,294.509,	Azote,
Vapour of water.........	2,219	428.726,	Vapour,
Carbonic acid gas	222	105.130,	Carbonic acid,
Total..........	887,528	274,345.111,	Of all kinds.

This is, in bulk, 25,607¼ imperial pints, or 57 hogsheads, 1 gallon, and 7¼ pints, and in weight 571½ ounces and 25 grains.

455. Now, although the air expired, in consequence of its recomposition, may have undergone changes in bulk, yet it seems agreeable to all analogy to suppose that its weight will remain the same as the weight inhaled. This, however, is not asserted as a truth, but only assumed, in order to show the result of such a theory.

456. Then the air expired in twenty-four hours will be as follows :—

	Bulk in cubic inches.	Weight in grains troy.
Given out undecomposed as before - - - - - - -	665,646	205,758.833
Recomposed carbonic acid gas - - - - - - -	38,268	18,130.147
Azote liberated - - - -	165,927	50,027.405
Vapour of water as before	2,219	428.726
Total - -	872,060	274,345.111

weighing as before, but less in bulk by 446¼ pints: so that for every 100,000 inches expired there were inspired 101,774 cubic inches.

457. When from the weight of carbonic acid gas thus expired, viz.,	18,130.147
we deduct the small portion inhaled in solution with the air - - - - -	105.130
The remainder is - - - -	18,025.017

The constituent parts of which are, oxygen derived from the air - - -	13,109.104
And pure carbon derived from the blood being the difference - - -	4,915.913

Thus in the compass of twenty-four hours the blood has produced 10 ounces and 116 grains very nearly of pure carbon.

	Grains.
458. Now, from the oxygen consumed in twenty-four hours as above	15,757.913
Deduct the weight restored in the form of carbonic acid gas - - - -	13,109.104
The remainder must have been absorbed into the blood - - - - - -	2,648.809
But the weight of carbon given out being as above - - - - - - - - - -	4,915.913
There is still an excess given out weighing	2,267.104

459. Some azote, however, is absorbed into the blood (439) as well as the above ascertained quantity of oxygen.

The weight of azote so absorbed must be precisely - - - - - - - - - - - -	2,267.104
if the theory be true, that equal weights are expired and inspired. In which case, as the weight of the azote of the air inspired was, as shown above - - - - - - - - - -	52,294.509

While the azote expired could only have weighed - - - - - - - - - - -	50,027.405
The difference would have been absorbed - - - - - - - - - - - - - -	2,267.104

And thus the weight of carbon discharged by the blood is precisely compensated by the united weight of the oxygen and azote which it has absorbed.

460. Since it appears to be a general truth that one quarter of the air respired is decomposed, and that the volume of air continually present in the lungs is sufficient for that consumption of oxygen which is requisite in 160 seconds of time, *if that volume be*, as is apparent, 48 *times the quantity decomposed* out of a single respiration, no error in the quantity of oxygen consumed in the twenty-four hours, which we have assumed, will affect the time of 160 seconds. For there being $18 \times 60 \times 24$ respirations, and $60 \times 60 \times 24$ seconds of time in the twenty-four hours, the 48th part of the first, and the 160th part of the last product is equally the 540th part of the whole, whatever it may be.

461. But if the time in which a circuit of the blood is performed be, as is most evident, identical with the time in which the whole volume of air in the lungs is decomposed, and if such period of time were, as the old observers have assigned, 150

seconds, then it would follow that only 45 times the quantity of air decomposed at a breath is present in the lungs, amounting to 385¼ cubic inches, and that the whole blood in the body is 24 ounces less than on the supposition of 160 seconds, that is to say, only 360 ounces, or 22½ pounds avoirdupois. Because the 45th part of $18 \times 60 \times 24$ is the same as the 150th part of $60 \times 60 \times 24$; in each it is the 567th part of the whole.

462. From the whole of these observations and calculations the following general results are deduced:—

1. The volume of air ordinarily present in the lungs is very nearly twelve pints (449).
2. The volume of air received by the lungs at an ordinary inspiration is one pint (422).
3. The volume of air expelled from the lungs at an ordinary expiration is a very little less than one pint (456).
4. Of the volume of air received by the lungs at one inspiration, only one-fourth part is decomposed at one action of the heart (447).
5. The fourth part of the volume of air received by the lungs at one inspiration, and decomposed at one action of the heart, is so decomposed in the five-sixth parts of one second of time (429.3).
6. The time in which a circuit of blood is performed is identical with the time in which the whole volume of air in the lungs is decomposed (461).

7. The whole volume of air decomposed in twenty-four hours is 221,882 cubic inches, exactly 540 times the volume of the contents of the lungs; 160 seconds being also exactly the 540th part of the number of seconds in twenty-four hours (450).

8. The quantity of the blood that flows to the lungs to be acted upon by the air at one action of the heart is two ounces (425).

9. This quantity of blood is acted upon by the air in the five-sixth parts of one second of time (429.3).

10. One circuit of the blood is performed in 160 seconds of time. Three circuits are performed every eight minutes; 540 circuits are performed in the twenty-four hours (453).

11. The quantity of blood in the whole body of the human adult is 24 pounds avoirdupois, or 20 pints imperial measure (428).

12. In the space of twenty-four hours, 57 hogsheads of air flow to the lungs (429.7).

13. In the same space of time 24 hogsheads of blood are presented in the lungs to this quantity of air (424.10).

14. In the mutual action that takes place between these quantities of air and blood, the air loses 15,757.9131 grains, or 328¼ ounces of oxygen, and the blood 10 ounces and 116 grains of carbon (445).

15. The blood, while circulating through the lungs, permanently retains and carries into the

system—of oxygen, 2,648,809 grains; and of azote, 2,267,104 grains (458).

16. The ultimate results are two :—

1st. While the chemical composition of the blood is essentially changed, its weight amidst all these complicated actions is maintained steadily the same; for the weight of carbon which is discharged by the blood is precisely compensated by the united weight of the oxygen and azote which it absorbs (459).

2ndly. The distribution of quantities is universally by proportions or multiples. Thus, of the air inspired, one measure is decomposed and three measures are returned unchanged: of the air decomposed at a single inspiration, there are always in store in the lungs precisely forty-eight measures; and so on in many other cases. The proportions are not arithmetical, but geometrical. When we compare arithmetical quantities with each other, we say that one quantity is by so much greater than another; when we compare geometrical quantities, we say that one quantity is so many times greater than another. From this adoption in the distribution of quantities of geometrical proportions it results that whatever be the size of the animal the ratios remain uniformly the same, and that thus one and the same law is adapted to the vital agencies of living beings under every possible diversity of magnitude and circumstance.

463. Such are the interesting and important

properties and relations deducible from the phenomena of respiration. The disappearance of oxygen and azote from the air inspired, and the replacement of the oxygen that disappears by the production of carbonic acid, and of the azote by the exhalation of azote, in which, as we have seen, the great changes wrought by respiration on the air consist, are essentially the same in all animals, whatever the medium breathed, and whatever the rank of the animal in the scale of organization. In all, the proportion of the oxygen of the inspired air is diminished;—in all, carbonic acid gas is produced. Comparing, then, the ultimate result of the function of respiration in the two great classes of living beings, it follows that the plant and the animal produce directly opposite changes in the chemical constitution of the air. The carbonic acid produced by the animal is decomposed by the plant, which retains the carbon in its own system and returns the oxygen to the air. On the other hand, the oxygen evolved by the plant is absorbed by the animal, which in its turn exhales carbonic acid for the re-absorption of the plant.

464. Thus the two great classes of organized beings renovate the air for each other, and maintain it in a state of perpetual purity. The plant, it is true, absorbs oxygen during the night as well as the animal; but the quantity which it gives off in the day more than compensates for that which it abstracts in the absence of light. This interesting

fact has been recently established by an extended series of experiments instituted by Professor Daubeney* for the express purpose of investigating this point.

465. From the general tenor of these experiments, it appears that, in fine weather and as long as the plant is healthy, it adds to the atmosphere an amount of oxygen not only sufficient to compensate for the quantity it abstracts in the absence of light, but to counterpoise the effects produced by the respiration of the whole animal kingdom. The result of one of these experiments will convey some conception of the amount of oxygen evolved. A quantity of leaves about fifty in number were enclosed in a jar of air; the surface of all the leaves taken together was calculated at about three hundred square inches; by the action of these leaves on the carbonic acid introduced into the jar, there was added to the air contained in it no less than twenty-six cubic inches of oxygen. As there was reason to conclude that the evolution of oxygen, in the circumstances under which this experiment was performed, was considerably less than it would have been in the open air, several plants were introduced into the same jar of air in pretty quick

* On the Action of Leaves upon Plants, and of Plants upon the Atmosphere, by Charles Daubeney, M.D. F.R.S., Professor of Chemistry and Botany in the University of Oxford. Philosophical Transactions of the Royal Society of London, for the year 1836. Part I.

succession: the amount of oxygen now evolved was increased from twenty-one to thirty-nine per cent., and probably had not even then attained the limit to which the increase of this constituent might have been brought. From the proportions of the constituent elements of carbonic acid gas (442) it necessarily follows that, by the mere process of decomposition, out of every eleven grains of carbonic acid gas eight grains of oxygen must be liberated, three grains of carbon being retained by the plant, and consequently that eight grains of oxygen must be restored to the atmosphere, less only by so much as the plant itself may absorb. How great, then, must be the production of oxygen by an entire tree under favourable circumstances; that is, when animal respiration and animal putrefaction present to it an abundant supply of carbonic acid on which to act!

466. This influence, says Professor Daubeney, is not exerted exclusively by plants of any particular kind or description. I have found it alike in the monocotyledonous and dycotyledonous; in such as thrive in sunshine and those which prefer the shade; in the aquatic as well as in those of a more complicated organization. How low in the scale of vegetable life this power extends is not yet exactly ascertained; the point at which it stops is probably that at which there ceases to be leaves.

467. From the whole, then, it appears that the functions of the plant have a strict relation to those

of the animal; that the plant, created to afford subsistence to the animal, derives its nutriment from principles which the animal rejects as excrementitious, and that the vegetable and animal kingdoms are so beautifully adjusted, that the very existence of the plant depends upon its perpetual abstraction of that, without the removal of which the existence of the animal could not be maintained.

468. The changes produced upon the blood by the action of respiration are no less striking and important than those produced upon the air. The blood contained in the pulmonary artery, venous blood (fig. 140-7.), is of a purple or modena red colour: the moment the air transmitted to the blood by the bronchial tubes comes into contact with it, in the rete mirabile (fig. 140-10.), this purple blood is converted into blood of a bright scarlet colour. Precisely the same change is produced upon the blood by its contact with the air out of the body. If a clot of venous blood be introduced into a vessel of air, the clot speedily passes from a purple to a scarlet colour; and if the air contained in the vessel be analyzed, it is found that a large portion of its oxygen has disappeared, and that the oxygen is replaced by a proportionate quantity of carbonic acid. If the clot be exposed to pure oxygen, this change takes place more rapidly and to a greater extent; if to air containing no oxygen, no change of colour takes place.

469. Th: elements of the blood upon which a

portion of the air exerts its action are carbon and hydrogen. The oxygen of the air unites with the carbon of the blood and forms carbonic acid, and this gas is expelled from the system by the action of expiration. The constituent of the blood which affords carbon to the air would appear to be chiefly the red particles. The other portion of the oxygen of the air unites with the hydrogen which is expelled with the carbonic acid in the form of aqueous vapour The direct and immediate effect of the action of respiration upon the blood is then to free it from a quantity of carbon and hydrogen.

470. Physiologists are not agreed whether the union of the oxygen of the air with the carbon of the blood takes place in the lungs or in the system. Some experimentalists maintain that the oxygen which disappears from the air, and that which is contained in the carbonic acid, are exactly equivalent, so that no oxygen can be absorbed. According to this view, which has been clearly shown to be incorrect (459), the effect of respiration is merely to burn the carbon of the blood, just as the oxygen of the air burns wood in a common fire, the result of this combustion being the generation of carbonic acid, which is expelled from the system the moment it is formed.

471. The theory of Dr. Crawford is essentially the same, which supposes that venous blood contains a peculiar compound of carbon and hydrogen, termed *hydro-carbon*, the elements of which unite

in the lungs with the oxygen of the air, forming water with the one and carbonic acid with the other. Mr. Cooper, for many years past, has taught the same doctrine in his lectures, without any knowledge of the fact that Crawford had suggested a similar modification of his theory.

472. It is now established that more oxygen disappears than is accounted for by the amount of carbonic acid that is generated. The experiments of Dr. Edwards had already shown this in so decisive a manner that physiologists almost universally admitted it as an ascertained fact. The calculations of Mr. Finlaison, to whom the opinions of physiologists on this point were unknown, have now determined the precise amount of oxygen (444 *et seq.*), and the probable amount of azote (459) absorbed. By many physiologists it is supposed that the oxygen retained by the lungs, as long as it remains in this organ, enters only into a state of loose combination with the blood; that in this state of loose combination, it is carried from the lungs into the general system; and that it is only in the system that the union becomes intimate and complete. According to this view, the lungs are merely the portal by which the substances employed in respiration are received and discharged, the essential changes induced taking place in the system. That it is through the lungs that the oxygen required by the system is received, is an opinion founded on experiments no less

exact than decisive; it is in accordance with the most probable theory of the production and distribution of animal heat (chap. ix.); and the preponderance of evidence in its favour is so great that, in the present state of our knowledge, it may be considered as established; but it will appear hereafter that the lungs are by no means passive in the process, and that, physiologically considered, they as truly constitute a gland secreting carbonic acid gas as the liver is a gland secreting bile.

473. Such are the main facts which have been ascertained relative to respiration, as far as this function is performed by the lungs. But the liver is a respiratory organ as well as the lungs. It decarbonizes the blood. It carries on this process to such an extent, that some physiologists are of opinion that the liver is the chief organ by which the decarbonization of the blood is effected. The following considerations show that whatever be the relative amount of its action, the liver powerfully co-operates with the lungs in the performance of a respiratory function.

1. The liver, like the lungs, is a receptacle of venous blood; blood loaded with carbon. The great venous trunk which ramifies through the lungs is the pulmonary artery, containing all the blood which has finished its circuit through the system. The great venous trunk which ramifies through the liver is the vena portæ, containing all the blood which has finished its circuit through the

apparatus of digestion. The liver is a secreting organ, distinguished from every other secreting organ by elaborating its peculiar secretion from venous blood. Carbon is abstracted from the venous blood that flows through the lungs in the form of carbonic acid; carbon is abstracted from the venous blood that flows through the liver in the form of bile.

2. All aliment, but more especially vegetable food, contains a large portion of carbon, more it would appear than the lungs can evolve. The excess is secreted from the blood by the liver, in the form of resin, colouring matter, fatty matter, mucus, and the principal constituents of the bile. All these substances contain a large proportion of carbon. After accomplishing certain secondary purposes in the process of digestion, these biliary matters, loaded with carbon, are carried out of the system together with the non-nutrient portion of the aliment. In the decarbonizing process performed by the lungs and the liver, the chief difference would seem, then, to be in the mode in which the carbon that is separated is carried out of the system. In the lungs it is evolved, as has been stated, in union with oxygen in the form of carbonic acid; in the liver, in union with hydrogen in the form of resin and fatty matter.

3. Accordingly, in tracing the organization of the animal body from the commencement of the

scale, it is found that among the distinct and special organs that are formed, the liver is one of the very first. It would appear to be constructed as soon as the economy of the animal requires a higher degree of respiration than can be effected by the nearly homogeneous substance of which, very low down in the scale, the body is composed. Invariably through the whole animal series, the magnitude of the liver is in the inverse ratio to that of the lungs. The larger, the more perfectly developed the lungs, the smaller the liver; and conversely, the larger the liver the smaller and the less perfectly developed the lungs. This is so uniform that it may be considered as a law of the animal economy. In the highly organized warm-blooded animal, with its large lungs, divided into numerous lobes, and each lobe composed of minute vesicles respiring only air, the magnitude of the liver compared with that of the body is small. In the less highly organized animal of the same class, with its smaller and less perfectly developed lung, respiring partly air and partly water, the liver increases as the lung diminishes in size. In the reptile with its little vesicular lung, divided into large cells, the liver is proportionally of greater magnitude. In the fish which has no lung, but which respires by the less highly organized gill, and only in the medium of water, the proportionate size of the liver is still greater; but in the molluscous animal, in which the lung or the gill is still less perfectly developed, the bulk of the liver is prodigious.

4. In all animals the quantity of venous blood which is sent to the liver increases, as that transmitted to the lung diminishes. In the higher animal the great venous trunk which ramifies through the liver (the vena portæ) is formed by the veins of the stomach, intestines, spleen, and pancreas, which are the only organs that transmit their blood to the liver. In the reptile, besides all these organs, the hind legs, the pelvis, the tail, the intercostal veins forming the vena azygos and in some orders of this class, even the kidneys also send their blood to the liver; but in the fish, in addition to all the preceding organs, the apparatus of reproduction likewise transmits its blood to the liver. The very formation of the venous system in the different classes of animals seems thus to point to the liver as a compensating and supplementary organ to the lung.

5. The permanent organs in the lower animal are a type of the transitory forms through which the organs of the higher animal pass in the progress of their growth. Thus the liver of the human fœtus is of such a disproportionate size, as to approximate it closely to that of the fish or of the reptile. After the birth of the human embryo, respiration is effected in part by the lung; but before birth the lung is inactive, no air reaches it;

it contributes nothing to respiration; the decarbonizing action of the blood is accomplished, not by the lung, but by the liver; hence the prodigious bulk of the fœtal liver and its activity in the secretion of bile, and especially towards the latter months of pregnancy, when all the organs are greatly advanced in size and completeness.

6. Pathology confirms the evidence derived from comparative anatomy and physiology. When the function of the lung is interrupted by disease, the activity of the liver is increased. In inflammation of the lung (pneumonia); in the deposition of adventitious matter in the lung (tubercles), by which the air vesicles are compressed and obliterated, the lung loses the power of decarbonizing the blood in proportion to the extent and severity of the disease with which it is affected. In this case the secretion of bile is increased. In diseases of the heart the liver is enlarged. In the morbus cæruleus (516) the liver retains through life its fœtal state of disproportion.

7. In the last place, there is a striking illustration of the respiratory action of the liver, in the vicarious office which it performs for the lung, during the heat of summer in cold, and all the year round in hot climates. In the heat of summer, and more especially in the intense and constant heat of a warm climate, in consequence of the rarefaction of the air, respiration by the lung is less active and efficient than in the winter of the cold

climate. During the exposure of the body to this long-continued heat, there is a tendency to the accumulation of carbon in the blood. An actual accumulation is prevented, by an increased activity in the secretion of bile, to which the liver is stimulated by the heat. In order to obtain the material for the formation of this unusual quantity of bile, it abstracts carbon largely from the blood; to this extent it compensates for the diminished efficiency of the lung, and thus removes through the vena portæ that superfluous carbon which would otherwise have been excreted through the pulmonary artery.

474. Taking life in its most extended sense, as comprehending both the circles it includes, the organic and the animal (vol. i. chap. 2), it may be said to have three great centres, of which two relate to the organic, and the third to the animal life (vol. i. chap. 2). The two centres which relate to the organic life are the systems of respiration and circulation; the third, which relates to the animal life, is the nervous system. Of the organic life, the lungs and the heart are the primary seats; of the animal, the brain and the spinal cord. Between each the bond of union is so close, that any lesion of the one influences the other, and neither can exist without the support of all. They form a triple chain, the breaking of a single link of which destroys the whole.

475. But of these three great centres of life,

upon which all the other vital phenomena depend, the most essential is respiration; hence, to consider the relation of this function to the others, is to take the most comprehensive view of the uses which respiration serves in the economy.

476. The first and most important use of the function of respiration is to maintain the action of the organs of the animal life. It has been shown (vol. i. chap. 2) that the organic is subservient to the animal life, and that to build up the apparatus of the latter, and to maintain it in a condition fit for performing its functions, is the final end of the former. The direct and the immediate effect of the suspension of respiration is the abolition of both functions of the animal life—sensation and voluntary motion. If a ligature be placed around the trachea of a living animal so as completely to exclude all access of air to the lungs, and if the carotid artery be then opened, and the blood allowed to flow, the bright scarlet-coloured blood contained in the artery is observed gradually to change to a purple hue. The exact point of time at which this change begins may be noted. It is seen to assume a darker tinge at the end of half a minute; at the end of one minute its colour is still darker, and at the end of one minute and a half, or at most two minutes (426), it is no longer possible to distinguish it from venous blood. As soon as this change of colour begins to be visible the animal becomes uneasy; his agitation increases

as the colour deepens; and when it becomes completely dark, that instant the animal falls down insensible. If in this state of insensibility air be readmitted to the lungs, the dark colour of the blood rapidly changes to a bright scarlet, and instantly sensation and consciousness return. But if, on the contrary, the exclusion of the air be continued for the space of three minutes from the first closing of the trachea, the animal not only remains to all appearance dead, but in general no means are capable of recovering him from the state of insensibility; and if the exclusion of the air be protracted to four minutes, apparent passes into real death, and recovery is no longer possible. It follows that one of the conditions essential to the exercise of the function of the brain is, that this organ receive a due supply of arterial blood.

477. The second use of the function of respiration is to afford blood capable of maintaining the muscles in a condition fit for the performance of their peculiar office, that of contractility. The closure of the trachea not only abolishes sensation, but the power of voluntary motion: sensation and motion are lost at once: on the re-admission of air to the lungs, both functions are regained at once: it follows that the process of respiration is as essential to the action of the muscle as to that of the brain. "By arterial blood," says Young, "the muscles are furnished with a store of that unknown principle by which they are rendered

capable of contracting." "The oxygen absorbed by the blood," says Spalanzani, "unites with the muscular fibres and endows them with their contractility." It is more correct to say, respiration takes carbon from the blood and gives it oxygen, and by this means endows the blood with the power of maintaining the contractility of the muscular fibre.

478. But respiration is as essential to the action of the organs of the organic life as to those of the animal. In a short time after the respiration ceases; the circulation stops. When the blood is no longer changed in the lungs, it soon loses all power of motion in the system; because venous blood paralyses the muscular fibres of the heart as of thé arm. When the left ventricle of the heart sends out venous blood to the system, it propels it into its own nutrient arteries, as well as into the other arteries of the body; into the coronary arteries, as well as into the other branches of the aorta; the heart loses its contractility, for the same reason as every muscle under the like privation; because venous instead of arterial blood flows in its nutrient arteries; and the circulation stops when the heart is no longer contractile, because the engine is destroyed that works the current.

479. Venous blood consists of chyle, the nutritive fluid formed from the aliment; of lymph, a fluid composed of organic particles, which having

already formed an actual part of the solid structures of the body, are now returning to the lungs to receive a higher elaboration; and of blood which, having completed its circuit through the system, and there given off its nutrient and received excrementitious matter, is now returning to the lungs for depuration and renovation. These commingled fluids, on parting in the lungs with carbonic acid and water, and on receiving in return oxygen and azote, are converted into arterial blood; that is, blood more coagulable than venous, and richer in albumen, fibrin, and red particles, the proximate organic principles of all animal structures. The rich and pure stream thus formed is sent out to the various tissues and organs, from which, as it flows to them, they abstract the materials adapted to their own peculiar form, composition, and vital endowments. By the reception of these materials the organs are rendered capable of performing the vital actions which it is their office to accomplish. And thus the processes of digestion, absorption, secretion, nutrition, formation, reproduction, all the processes included in the great organic circle, no less than muscular action and nervous energy, depend on receiving a due supply of arterial blood. All these actions, like the faculties of the animal life, cease totally and for ever in a few minutes after the formation of this vital fluid has been stopped by the suspension of respiration.

480. In the last place, the depurating process

effected by respiration is necessary to prevent the decomposition of the blood, and eventually that of the body. The first step in the spontaneous decomposition of animal matter consists in the loss of a portion of its carbon, which, uniting with the oxygen of the atmosphere, forms carbonic acid; precisely the same thing that takes place in the process of respiration. The bodies of all animals, of worms, insects, fishes, birds, and mammalia, deoxidate the air and load it with carbonic acid after death, some of them nearly as much as during life; and this before any visible marks of decomposition can be traced. It is probable that the cause which more immediately operates in preventing the decomposition of the body is the abstraction of a part of the carbon of the blood; that were these carbonaceous particles allowed to accumulate, they would produce a tendency to decomposition, which would terminate in complete disorganization; and consequently, that one main object of the process of respiration is to afford blood not only capable of nourishing and sustaining the organs, but of maintaining their integrity, by removing noxious matter, the presence of which would subvert their composition and lead to their entire decomposition.

481. The ultimate object of respiration, then, is to prepare and to preserve in a state of purity a fluid capable of affording to all the parts of the body the materials necessary to maintain their

vital endowments. By the exhalation of oxygen and water, and the absorption of carbon, under the agency of light, the plant elaborates such a fluid from its nutritive sap, and out of this elaborated sap forms terniary combinations, the organic elements of all vegetable solids. By the absorption of oxygen and azote, and the exhalation of carbonic acid and water, probably under the influence of electricity, conducted and regulated by the nervous system, the animal elaborates such a fluid from its aliment, and out of this elaborated fluid forms quaternary combinations, albumen, and fibrin, the organic elements of all animal solids.

CHAPTER IX.

Of the temperature of living bodies—Temperature of plants—Power of plants to resist cold and endure heat—Power of generating heat—Temperature of animals—Warm-blooded and cold-blooded animals—Temperature of the higher animals—Temperature of the different parts of the animal body—Temperature of the human body—Power of maintaining that temperature at a fixed point whether in intense cold or intense heat—Experiments which prove that this power is a vital power—Evidence that the power of generating heat is connected with the function of respiration—Analogy between respiration and combustion—Phenomena connected with the functions of the animal body, which prove that its power of generating heat is proportionate to the extent of its respiration—Theory of the production of animal heat—Influence of the nervous system in maintaining and regulating the process—Means by which cold is generated, and the temperature of the body kept at its own natural standard during exposure to an elevated temperature.

482. Closely connected with the function of respiration, is the power which all living beings possess of resisting within a certain range the influence of external temperature. The plant is warmer than the surrounding air in winter, and colder in summer. A thermometer placed at the

bottom of a hole bored into the centre of a living tree, precaution being taken to keep off as much as possible all external influence either of heat or cold, does not rise and fall according to the changes of external temperature; but rises when the external air is cold, and falls when it is warm. Thus, in a cold day in spring, the wind being north, at six o'clock in the evening, the temperature of the external air being 47°, that of a tree was 55°. On another cold day in the same month, there being snow and hail, and the wind in the north-east, at six o'clock in the evening, the external temperature being 39°, that of the tree was 45°. On the contrary, in one experiment, when the temperature of the air was 57½°, that of the tree was only 55°; and when the temperature of the air was 62°, that of the tree was 56°.

483. These experiments afford an explanation of circumstances familiar to common observation. Every one has noticed that the snow which falls on grass and trees melts rapidly, while that on the adjoining gravel walks often remains a long time unthawed. Moist dead sticks are constantly found frozen hard in the same garden with tender growing twigs, which are not in the least degree affected by the frost. Every winter in our own climate tender herbaceous plants resist degrees of cold which freeze large bodies of water.

484. But the colder, and the warmer the climate, the more strikingly does the plant exemplify

the power with which it is endowed of resisting external temperature. In the northern parts of America the temperature is often 50° below zero; yet, though exposed to this intense degree of cold, the spruce fir, the birch, the juniper, &c. preserve their vitality uninjured. From numerous experiments which have been performed expressly with a view to ascertain this point, it is found that a plant which has been once frozen is invariably dead when thawed. It is also proved by direct experiment, that if the sap be removed from its proper vessels, it freezes at 32°, the ordinary freezing point. In the northern parts of America, then, the plant must preserve in its living vessels its sap from freezing, when exposed to a temperature of 50° below zero; which sap out of these vessels would congeal at the ordinary freezing point; that is, the plant of this climate is endowed with the power of resisting a degree of cold ranging from the ordinary freezing point to 50° below zero; a property which can be referred only to a vital power, by the operation of which the plant generates within itself a degree of heat sufficient to counteract the external cold.

485. The opposite faculty of resisting the influence of external heat is exemplified by the trees and shrubs of tropical climates, often surrounded by a temperature of 104°, which they resist just as the plant of the northern clime resists the intense degrees of cold to which it is exposed.

486. That the plant is endowed with the power of generating heat is demonstrated by the phenomena which attend the performance of some of its vital processes, such as those of germination and flowering. During the germination of barley, the thermometer was observed to rise in the course of one night to 102°. The bulb of a thermometer applied to the surface of the spadix of an arum maculatum, indicated a temperature 7° higher than that of the external air; but in an arum cordifolium, at the Isle of France, a thermometer placed in the centre of five spadixes stood at 111°; and in the centre of twelve at 121°, though the temperature of the external air was only 66°.

487. Animals indicate in a still more striking degree the power of generating heat. The lower the animal in the scale of organization, indeed, the nearer it approaches to the plant in the comparative feebleness of this function. The heat of worms, insects, crustacea, mollusca, fishes, and amphibia, is commonly only two or three degrees above that of the medium in which they are immersed. Absolutely colder than the higher animals, they are at the same time incapable of resisting any considerable changes in the temperature of the surrounding medium, whether from heat to cold or from cold to heat. The higher animals, on the contrary, maintain their heat steadily at a fixed point, or very nearly at a fixed point, however the temperature of the surrounding medium may

change. Hence animals are divided into two great classes, the cold-blooded and the warm-blooded. The temperature of the cold-blooded is lower than that of the warm-blooded, and it varies with the heat of the surrounding medium ; the temperature of the warm-blooded is higher than that of the cold-blooded, and it remains nearly at the same fixed point, however the heat of the surrounding medium may change.

488. The temperature natural to the higher animals differs somewhat according to their class. The temperature of the bird is the highest, and is pretty uniformly about 103° or 104°; that of the mammiferous quadruped is 100 or 101°; that of the human species is 97° or 98°.

489. The temperature of the animal body is not precisely the same in every part of it. The ball of the thermometer introduced within the rectum of the dog stood at 100½; within the substance of the liver at 100¾; within the right ventricle of the heart at 101°, and within the cavity of the stomach at 101°. In the brain of the lamb it stood at 104°; in the rectum at 105°; in the right ventricle of the heart, and in the substance of the liver and of the lungs, at 106°; and in the left ventricle of the heart at 107°.

490. The temperature natural to the human body is 98°. When the human body is surrounded by an atmosphere at the temperature of 30°, it must have its heat rapidly extracted by the cold medium ;

yet the temperature of the body, however long it remain exposed to such a degree of cold, does not sink, but keeps steadily at its own standard. But animals which inhabit the polar regions are often exposed to a cold 40° below zero. The temperature of Melville Island is so low during five months of the year that mercury congeals, and the temperature is sometimes 46° below zero; yet the musk oxen, the rein deer, the white hares, the polar foxes, and the white bears which abound in it maintain their temperature steadily at their own natural standard.

491. The power which the higher animal possesses of resisting heat is still more remarkable than its power of resisting cold. On taking rabbits and guinea-pigs from the temperature of 50°, and introducing them very rapidly to the temperature of 90°; it was found that the animals acquired only two or three degrees of heat. How different the result when the cold-blooded animal is subjected to the same experiment! The temperature of the surrounding air being 45°, a thermometer introduced into the stomach of a frog rose to 49°. The frog being then put into an atmosphere made warm by heated water, and allowed to stay there twenty minutes, the thermometer on being now introduced into the stomach rose to 64°.

492. But the human body may be actually placed in a temperature of 60° above that of boiling water, not only without sustaining the slightest in-

jury, but without having its own temperature raised excepting by two or three degrees. The attention of physiologists was first directed to this curious fact by some remarkable circumstances related by the servants of a baker at Rochefoucault, who were in the habit of going into the heated ovens in order to prepare them for the reception of the loaves. In performing this service, the young women were sometimes exposed to a temperature as high as 278°. It was stated that they could endure this intense heat for twelve minutes, without any material inconvenience, provided they were careful not to touch the surface of the oven. Subsequently Drs. Fordyce, Blagden, and others, with a view to ascertain the exact facts, entered a chamber, heated to a temperature much above that of boiling water, and some of the phenomena observed during these experiments are highly curious.

493. In the first room entered by these experimentalists, the highest thermometer varied from 132° to 130°; the lowest stood at 119°. Dr. Fordyce having undressed in an adjoining cold chamber, went into the heat of 119°; in half a minute the water poured down in streams over his whole body, so as to keep that part of the floor where he stood constantly wet. Having remained here fifteen minutes, he went into the heat of 130°; at this time the heat of his body was 100°, and his pulse beat 126 times in a minute. While Dr. Fordyce stood in this situation a Florence

flask was brought in by his order, filled with water heated to 100°, and a dry cloth with which he wiped the surface of the flask quite dry; but it immediately became wet again, and streams of water poured down its sides, which continued till the heat of the water within had risen to 122°, when Dr. Fordyce went out of the room, after having remained fifteen minutes in a heat of 130°: just before he left the room his pulse made 129 beats in a minute; but the heat under his tongue and in his hand did not exceed 100°.

494. In a subsequent experiment the chamber was entered when the thermometer stood above 211°. The air heated to this degree, says Dr. Blagden, felt unpleasantly hot; but was very bearable. Our most uneasy feeling was a sense of scorching in the face and legs; our legs particularly suffered very much, by being exposed more fully than any other part to the body of the stove, heated red hot by the fire within. Our respiration was not at all affected; it became neither quick nor laborious; the only difference was a want of that refreshing sensation which accompanies a full inspiration of cool air. But the most striking effects proceeded from our power of preserving our natural temperature. Being now in a situation in which our bodies bore a very different relation to the surrounding atmosphere from that to which we had been accustomed, every moment presented a new phenomenon. Whenever we

breathed on a thermometer, the quicksilver sank several degrees. Every expiration, particularly if made with any degree of violence, gave a very pleasant impression of coolness to our nostrils, scorched before by the hot air rushing against them whenever we inspired. In the same manner our now cold breath agreeably cooled our fingers whenever it reached them. Upon touching my side, it felt cold like a corpse; and yet the actual heat of my body, tried under my tongue, and by applying closely the thermometer to my skin, was 98°, about a degree higher than its ordinary temperature. When the heat of the air began to approach the highest degree which this apparatus was capable of producing, our bodies in the room prevented it from rising any higher; and when it had been previously raised above that point, invariably sunk it. Every experiment furnished proofs of this. Mr. Banks and Dr. Solander each found that his single body was sufficient to sink the quicksilver very fast, when the room was brought nearly to its maximum of heat.

495. In a third series of experiments the temperature of the chamber was raised to the 260th degree. At this time, continues Dr. Blagden, I went into the room, with the addition to my common clothes of a pair of thick worsted stockings drawn over my shoes, and reaching some way above my knees. I also put on a pair of gloves, and held a cloth constantly between my face and

the stove (necessary precautions against the scorching of the red-hot iron). I remained eight minutes in this situation, frequently walking about to all the different parts of the room, but standing still most of the time in the coolest spot near the lowest thermometer. The air felt very hot, but by no means so as to give pain. I had no doubt of being able to bear a much greater heat; and all who went into the room were of the same opinion. I sweated, but not very profusely. For seven minutes my breathing remained perfectly good; but after that time, I began to feel an oppression in my lungs, attended with a sense of anxiety; which gradually increasing for the space of a minute, I thought it most prudent to end the experiment. My pulse, counted as soon as I came into the cool air, for the uneasy feeling rendered me incapable of examining it in the room, beat at the rate of 144 pulsations in a minute, which is more than double its ordinary quickness. In the course of this experiment, and others of the same kind by several of the gentlemen present, some circumstances occurred to us which had not been remarked before. The heat, as might have been expected, felt most intense when we were in motion; and on the same principle, a blast of the heated air from a pair of bellows was scarcely to be borne: the sensation in both these cases exactly resembled that felt in our nostrils on inspiration. It was observed that our breath did not feel cool to our fingers unless held very near the mouth; at a distance the cooling power of the breath did not sufficiently compensate the effect of putting the air in motion, especially when we breathed with force.

496. On going undressed into the room, the impression of the air was much more disagreeable than before; but in five or six minutes, a profuse sweat broke out, which instantly relieved me. During all the experiments of this day, whenever I tried the heat of my body, the thermometer always came very nearly to the same point (the ordinary standard), not even one degree of difference, as in our former experiments.

497. To prove that there was no fallacy in the degree of heat shown by the thermometer, but that the air which we breathed was capable of producing all the well-known effects of such heat on inanimate matter, we put some eggs and a beef steak upon a tin frame, placed near the standard thermometer, and farther distant from the stove than the wall. In about thirty minutes the eggs were taken out roasted quite hard. In about forty-seven minutes the steak was not only dressed, but almost dry. Another beef steak was rather overdone in thirty-three minutes. In the evening when the heat was still greater, we blew upon a third steak with the bellows, which produced a visible change on its surface, and hastened its

dressing; the greatest part of it was pretty well done in thirteen minutes.

498. The human body, then, may be exposed to a temperature 50° below zero, without having its own heat appreciably diminished; it may be exposed to a temperature 60° above that of boiling water, without having its own heat increased beyond two or three degrees; or, as appears from experiments subsequently performed expressly to ascertain this point, from three to five degrees. In the former case, the body must generate a degree of heat sufficient to compensate the great quantity of caloric which is every moment abstracted from it by the intensely-cold surrounding medium. In the latter case it must generate a degree of cold sufficient to counteract the great quantity of caloric which is every moment communicated to it by the intensely-hot surrounding medium.

499. Powers so wonderful and so opposite appeared to the physiologists of former times to be involved in such profound mystery, that they did not even attempt to investigate their nature, or trace their mode of operation; but satisfied themselves with referring them to some innate quality of the body, and with considering them as essential attributes of life. And difficulties connected with the subject still remain, which the present state of knowledge does not permit us wholly to surmount;

but we are able at least to refer these powers to their proper seat, and to trace some steps of the processes by which they produce results so wonderful and beautiful.

500. It is certain that whatever be the ultimate physical processes by which the generation of heat and the production of cold are effected in the animal body, the phenomena are dependent on the condition of life. No such phenomena take place excepting in living bodies. This is illustrated in a striking manner by a series of experiments performed by Mr. Hunter. A part of the living human body was immersed in water gradually made warmer and warmer from 100° to 118°; precisely the same part of the body, dead, was immersed in the same water, and both parts, the living and the dead, were continued in this heat for some minutes. The dead part raised the thermometer to 114°; the living part raised it to no higher than 102¼°. On applying the thermometer to the sides of the living part, the quicksilver immediately fell from 118° to 104°; on applying it close to the dead part, the thermometer did not fall above a single degree; the living part actually produced a cold space of water around it. Hence in bathing in water, whether colder or warmer than the heat of the body, the water soon acquires the same temperature with that of the body; and, consequently, in a large bath the patient should move from place to place, and in a

small one there should be a constant succession of water of the intended heat.

501. A fresh, that is, a living egg was put into cold water at about zero, frozen, and then allowed to thaw. By this process its vitality was destroyed, and consequently its power of resisting cold and heat lost. This thawed egg was next put into a cold mixture with an egg newly laid: the time required for freezing the fresh egg was seven minutes and a half longer than that required for freezing the thawed egg.

502. A new-laid egg was put into a cold atmosphere fluctuating between 17° and 15°; it took about half an hour to freeze; but when thawed and put into an atmosphere at 25° (10° warmer), it froze in half the time.

503. A fresh egg and one that had been frozen and thawed were put into a cold mixture at 15°; the thawed one soon came to 32°, and began to swell and congeal; the fresh one sunk to 29½, and in twenty-five minutes after the dead one, it rose to 32°, and began to swell and freeze.

504. The result of this experiment upon the fresh egg was similar to that of analogous experiments made upon the frog, eel, snail, &c. where life allowed the heat to be diminished 2° or 3° below the freezing point, and then resisted all further decrease; but the powers of life having been expended by this exertion, the parts then froze like any other dead animal matter.

505. The heat of the bird is increased somewhat when it is prepared for incubation. Some eggs were taken from under a sitting hen whose temperature was 104°, at the time when the chick was about three-parts formed. A hole was broken in the shell and the bulb of a thermometer introduced; the quicksilver rose to 99½°; but in some eggs that were addled it was proved that their heat was not so high by two degrees, so that the life of the living egg assisted to support its own temperature.

506. These facts sufficiently show the dependence of the faculty of generating heat and of producing cold on the powers of life. But the processes by which, under the agency and control of the vital powers, these different results are effected, are various, and even opposite.

507. The power of generating heat is connected in the closest manner with the function of respiration, and is directly dependent upon it. The evidence of this is indubitable. For—

508. i. Respiration is combustion, and, like ordinary combustion, is attended with the production of heat. In ordinary combustion oxygen disappears, and a new compound is formed, consisting of oxygen combined with the combustible matter; that is, an oxidized body is generated. On burning a piece of iron wire in oxygen, the oxygen disappears, and the iron increases in weight. The oxygen combines with the iron;

forming a new product, oxide of iron, and the weight of this new substance is found on examination to be exactly equal to the weight of the wire originally employed, added to the quantity of oxygen which has disappeared.

509. It is precisely the same in respiration. In this process oxygen combines with combustible matter, carbon: the oxygen disappears, and a new body, carbonic acid, is generated.

510. ii. One phenomenon which invariably accompanies the combination of oxygen with combustible matter is the extrication of heat. Whenever a substance passes from a rarer into a denser state; when, for example, a gas is converted into a liquid or solid, or when a liquid solidifies, heat is evolved; because, according to the ordinary theory of combustion, the denser substance has a less capacity for caloric than the rarer, and consequently in passing from a rare into a dense state, a quantity of caloric previously combined or latent within it is set free. The combined or latent caloric contained in a body is termed its specific caloric; the caloric which is evolved on its change of state is named free or sensible caloric.

511. The combination of oxygen with carbon, as in the combination of oxygen with combustible matter in every other instance, must be attended with the evolution of heat. Though the product of the combustion, in the present case, be a gaseous body, carbonic acid, still, according to the ordinary

theory of combustion, carbonic acid has less specific caloric, or less capacity for caloric, than oxygen; and therefore in combining with carbon, a portion of its specific caloric becomes free or sensible, that is, heat is evolved. But whatever theory of combustion be adopted, the fact is certain, that whenever oxygen combines with carbon to form carbonic acid, heat is evolved; not only in the rapid union which takes place in ordinary combustion, but also in the slow combination which occurs in fermentation, putrefaction, and germination; in the latter of which processes, as in the malting of barley, the temperature rises as high as 10°. The union of oxygen with carbon in the lungs during respiration must therefore necessarily produce heat, just as it does in a charcoal fire, or in any other natural process in which this combination takes place.

512. iii. Numerous phenomena connected with the animal body show that its temperature is in strict proportion to the quantity of oxygen which is consumed in respiration, and to the quantity of carbonic acid which is formed by the union of oxygen and carbon during the process.

513. In all animals whose respiratory organs are so constructed, that the consumption of oxygen and the consequent generation of carbonic acid is minute in quantity, the production of heat is proportionably small. It has been shown (337 *et seq.*), that in almost the entire class of the

invertebrata, the respiratory apparatus is comparatively minute and imperfect; accordingly, in these animals the power of generating heat is at the minimum. In the fish, though the respiratory apparatus be large, and though all the blood of the body circulate through it (345 *et seq.*), yet only a small quantity of air is brought into contact with the respiratory organ, merely the air contained in water. In the reptile, though it possess a true and proper lung, and respire air, yet only one half of the blood of its body circulates through the comparatively small, imperfectly divided, and simply constructed air bag, which constitutes its respiratory organ (354). Hence, the striking contrast exhibited between the temperature of these cold-blooded creatures and that of the mammiferous quadruped, whose lung, comparatively large, and composed of innumerable minute and closely-set air vesicles (fig. CXXXIV. and CXXXV.), presents to the air an immense extent of surface (370), and the whole mass of whose blood incessantly traversing this surface, comes at every point into contact with the air (399).

514. In the various tribes of warm-blooded animals, the elevation and uniformity of the temperature is strictly proportionate to the comparative magnitude of the lungs; to the complexity of their structure; to the minuteness and number of the air vesicles; and, consequently, to the quantity of oxygen consumed, and of carbonic acid generated.

515. In all animals with red blood there is a strict relation between the temperature of the body and the lightness or depth of the colour of the blood; invariably the deeper the colour, the higher the temperature. Thus, the blood of the fish and of the reptile is of a light, and that of the bird of an intense red colour. It has been shown (229) that the lightness or deepness of the colour of the blood depends on the quantity of red particles which it contains, and the chemical action between the air and the blood is carried on chiefly through the medium of the red particles.

516. Even in the same animal, the temperature differs at different times, according to the energy with which the process of respiration is carried on. When the circulation of the blood is sluggish and the respiration slow and feeble, the quantity of oxygen consumed is small, and the temperature low; when, on the contrary, the circulation is rapid, and the respiration energetic, the quantity of oxygen consumed is large, and the temperature proportionably high. Whatever diminishes the quantity of air that flows to the lungs, and the quantity of blood that circulates through them, diminishes the temperature. Malformation of the heart, in consequence of which a quantity of blood is sent to the system without passing through the lungs, as in the individuals termed Ceruleans: disease of the lungs, by which the access of air to the air vesicles is obstructed, as in asthma, are

morbid states invariably attended with a diminution of the temperature.

517. When a warm-blooded animal is placed in an elevated temperature, its consumption of oxygen is comparatively small; when it is placed in a cold atmosphere, and the production of a large quantity of heat is necessary to maintain its temperature at its natural standard, its consumption of oxygen is proportionably large; accordingly, it is established by direct experiment that the same animal consumes a much larger quantity of oxygen in winter than in summer.

518. Due allowance being made for the difference in their bulk, young animals consume less oxygen than adults; and they have a less power of generating heat. Different species of young animals differ from each other in their power of generating heat, and the closest relation is observable between the difference in their power of consuming oxygen and that of generating heat. Puppies and kittens require so small a quantity of oxygen for supporting life, that they may be wholly deprived of this gas for twenty minutes, without material injury, while adult animals of the same species perish when deprived of it only for four minutes. As long as these young creatures retain the power of sustaining life for so protracted a period without oxygen, they are wholly incapable of maintaining their own temperature; on free exposure to air, even in summer, the heat of their

body sinks rapidly, and if this exposure be continued long, they perish of cold. In like manner, young sparrows and other birds which are naked when hatched, consume little oxygen, and are incapable of maintaining their temperature; but can support life when deprived of oxygen much longer than adult birds of the same species; while young partridges which are able to retain their own temperature at the period of quitting the shell, die when deprived of oxygen as rapidly as the adult bird.

519. The state of hybernation illustrates in the same striking manner the relation between respiration and the generation of heat. One of the most remarkable phenomena connected with this curious state, is the reduction, sometimes even the apparent suspension, of respiration; and in all cases of hybernation, the respiratory function is performed in a feeble manner, and only at distant intervals. Exactly in proportion to the diminution of the respiration, is the reduction of the power of generating heat; so that when the state of hybernation is established, the temperature of the external parts of the body sinks nearly to that of the surrounding medium; while the internal parts, the blood, and the vital organs are only a degree or two higher. In experiments made to reduce an hybernating animal to a torpid state by cold artificially produced, De Saissy found that he could not bring on the state of hybernation by the

reduction of temperature alone, without also constraining the respiration.

520. These and other analogous facts abundantly establish the relation between the function of respiration and that of calorification, and lead to the general conclusion that the generation of animal heat is in the direct ratio of the quantity of air and blood which are brought into contact, and which act on each other in a given time. Yet an attempt has recently been made by an ingenious physiologist* to disturb this induction, and to show that the production of animal heat is not in the direct ratio of the quantity of oxygen inhaled, but in the inverse ratio of the quantity of blood exposed to this principle. This position is maintained on the following grounds:—

521. Inspiration favours the flow of blood to the lungs; expiration retards it: consequently, if from any causes the inspirations preponderate in number and proportion over the expirations, a greater quantity of blood than usual will be accumulated in the lungs. There are conditions of the system in which this preponderance of the inspirations actually takes place; when the mind is under the influence of certain emotions, for example, as when it is depressed by anxiety and fear. In this state the inspirations are more frequent

* An Experimental Inquiry into the Laws which regulate the Phenomena of Organic and Animal Life. By George Calvert Holland, M.D.

and more complete than the expirations; it is a state of continual sighing. In like manner, in certain diseases, such as asthma, the inspirations greatly preponderate both in frequency and energy over the expirations. In such conditions of the system the blood accumulates in preternatural quantity in all the internal organs; but more especially in the lungs; and two consequences follow: first, there is a remarkable diminution in the energy of all the vital actions; and secondly there is a proportionate diminution in the production of animal heat.

522. On the contrary, as it is the effect of inspiration to facilitate the motion of the blood through the lungs, so it is the effect of expiration to retard it; hence, when the expirations preponderate the opposite state of the system is induced; all the vital actions are performed with increased energy; the heart beats with unusual vigor; the pulse becomes quick and strong; a larger quantity of blood is determined to the surface of the body, and this excited state of the system is always attended with an augmentation of the temperature.

523. As in the first state there is a greater and in the second a smaller quantity of blood than natural contained in the lungs, the inference deduced by Dr. Holland is, that the production of animal heat is in the inverse ratio of the quantity of blood exposed to oxygen. But this inference is neither logical nor sound.

524. If, as a comparison of all the phenomena of respiration exhibited throughout the entire range of the animal kingdom, shows the production of animal heat to be in the direct ratio of the quantities of air and blood which are brought into contact, and which re-act on each other, every phenomenon of respiration must be in harmony with this law, and, accordingly, when really understood, it is found to be so.

525. Inspiration, by the dilatation of the thorax, and consequently of the lungs incident to that action, is favorable to the flow of blood to the lungs. But it is only a certain degree of dilatation of the lungs that is favorable to the flow of blood through them (407 *et seq.*). If the dilatation be carried beyond a certain point, the quantity of blood transmitted through the pulmonary tissue is diminished (406); if the dilatation be carried farther, the transmission of the blood may be wholly stopped (417). The quantity of the blood which flows to the lungs, and the quantity which circulates through them, are not then identical. So large a quantity may flow to them as to impede or retard or wholly stop the pulmonary circulation. In proportion to the accumulation of blood in the lung must necessarily be the distension of the pulmonary tissue; in that proportion the lung must be approximated to its condition in the experiment in which it was distended with water (417), when it did not transmit a single particle of blood. Further,

in proportion to the preternatural distension of the pulmonary tissue with blood must be the exclusion of air from the air vesicles for the lungs can contain only a certain quantity of blood and air (418.3), so that the blood can preponderate only by the exclusion of the air.

526. In those states of the system, then, in which the preponderance of the inspirations induces a preternatural accumulation of blood in the lungs, the production of animal heat is diminished for a two-fold reason; first, because the distension of the pulmonary tissue with blood retards the pulmonary circulation, and proportionally lessens the quantity of blood which is brought into contact with the air; and, secondly, because the distended blood-vessels compress the air vesicles, and so diminish the quantity of air which is brought into contact with the blood.

527. It follows that the diminution of temperature which takes place in this condition of the system is not because the production of animal heat is in the inverse ratio of the quantity of blood which is exposed to oxygen; but because from a twofold operation there is a diminution of the quantity of blood and of oxygen which are brought into contact.

528. The reason is equally obvious why there is an increase of the temperature in those conditions of the system in which the expirations preponderate over the inspirations. Expiration,

it is true, somewhat retards the circulation of the blood through the lungs, but the preponderance of this respiratory action does not raise the temperature by the retardation of the flow of blood through the lungs, and the consequent diminution of the quantity transmitted in a given time; for though expiration somewhat retards the circulation of the blood through the branches of the pulmonary artery, it promotes its circulation through the branches of the pulmonary veins (fig. CXL. 10). It is indeed by the action of expiration that the aerated blood is transmitted from the lungs to the left heart to be sent out renovated to the system. Expiration has no influence whatever over the aeration of the blood. Before the action of expiration takes place, the blood is already aerated. The office of expiration is to remove from the system the air which has served for respiration, and to transmit to the system the blood which has been subjected to respiration. Consequently, in those states of the system in which the expirations preponderate, the temperature is increased, not because the expiratory actions, by lessening the quantity of blood in the lungs, diminish the quantity exposed to oxygen, but because they transmit to the system oxygenated blood as rapidly as it is formed, that is, blood which either produces animal heat in the act of its formation, or which generates it as it flows through the system.

529. These conditions establish the conclusion deduced, as has been stated, from the comparison of the phenomena of respiration exhibited throughout the entire range of the animal kingdom. But if the production of animal heat be really the result of combustion, if that combustion take place in the lung, and if the lung be thus the focus whence the heat radiates to every other part of the body, why is not the heat of this organ and of the parts in its immediate neighbourhood higher than the temperature of the rest of the body? Some of the internal organs are indeed a degree or two hotter than the general mass of the circulating blood (469), and among these the lung is admitted to rank perhaps the very highest. But how can a quantity of caloric sufficient to maintain the heat of the body in a temperature of forty degrees below zero radiate from an organ the temperature of which is only two or three degrees above that of the body itself? It is estimated that, in every minute, during the calm respiration of a healthy man of ordinary stature, 26·6 cubic inches of carbonic acid, at the temperature of 50° Fahr. are emitted, and that an equal volume of oxygen is withdrawn from the atmosphere. From these data it is calculated that, in an interval of twenty-four hours, not less than eleven ounces of carbon are consumed. Why is the lung, the seat of this combustion, not only not greatly warmer

than any other organ; but why is it not even consumed by the fire which is thus incessantly burning within it?

530. It has been shown (468 and 469) that when the carbon of the blood unites in the lung with the oxygen of the air, the nature of the blood, in consequence of the abstraction of carbon, undergoes an essential change, passing from venous into arterial. By an elaborate series of experiments, conducted with extraordinary care and skill, it would appear that arterial has a greater capacity for caloric than venous blood, in the proportion of 114·5 to 100. In consequence of this difference n the constitution of the two kinds of blood, the heat generated in the lung by the combustion of carbon, instead of being evolved or becoming sensible (510. ii.), and so raising the temperature of the organ, goes to satisfy the increased capacity for caloric of arterial blood, is spent, not in rendering the fluid sensibly warmer, but in augmenting its specific caloric (510. ii.). Arterial blood is not increased in temperature,* but with its absolute

* It is not a perfectly accurate statement that the temperature of venous and arterial blood is precisely the same. The latest and best experiments concur in showing that arterial blood, at least in the heart and the great arterial trunks, is one or two degrees warmer than venous blood. The weight of evidence from experiment is also in favour of the opinion, that the different parts of the body are *somewhat* less warm as they recede from the lungs and heart; but the difference is so slight that it may be disregarded in the general argument.

quantity of caloric augmented, flows from the lung to the left heart (fig. CXL. 10), and thence to the system (fig. CXL. 6). In the system, in every organ, at every point of the component tissue of every organ and at every moment of time, the blood repasses from the arterial to the venous state: by this transition its capacity for heat is diminished; the venous cannot retain in it the same quantity of caloric as the arterial blood, consequently a portion of caloric is extricated; that which was latent becomes sensible, and caloric being set free the temperature is raised. In this process the lung is not burnt, it is only rendered just sensibly warmer than any other part of the body, though it be the organ by which the whole mass of blood receives its caloric, because it is only in the capillary part of the systemic circulation, when the arterial blood again passes into the venous state, that the caloric acquired is liberated. In this manner, gently, steadily, uninterruptedly, an abundant, unceasing, and equable current of heat is distributed to every part and particle of the system.

531. Such is the celebrated theory of animal heat suggested by Dr. Crawford, of which it has been justly said, that it affords one of the most beautiful specimens of the application of physical and chemical reasoning to the animal economy that has ever been presented to the world.

532. The main position on which this theory

rests—that arterial possesses a greater capacity for caloric than venous blood—professes to be founded on experiments which, though of a delicate and complex nature, are nevertheless uniform and decisive in their results. In consequence of their extreme interest and importance, these experiments have been subjected, by different physiologists, to rigid examination, with a somewhat conflicting result. The greater number of experimentalists maintain that Crawford's experiments are correct in all the essential points, and that the objections which have been urged against them do not really affect them; while others are of opinion that, even although it must, upon the whole, be admitted that the specific heat of arterial is greater than that of venous blood; yet that the excess is so small as to be inadequate to account for the effects attributed to it. Dr. Davy's experiments, which of all that have been instituted are generally conceived to be the most unfavourable to the theory of Crawford, do not afford uniform results. Three experiments out of four indicate a greater capacity in arterial than in venous blood; in those in which the experimentalist himself places the most confidence, in the relative proportion of 913 to 903; while, according to Crawford, the relative proportion is 114·5 to 100.

533. But when this subject is closely considered, the discrepancy in question turns out to be of no real consequence. There is a modification

of the theory, which removes every difficulty, and dispenses with the necessity of any regard whatever to the point in dispute.

534. It has been shown (444 *et seq.*), that during the process of respiration more oxygen disappears than is accounted for by the carbonic acid that is generated; that this excess of oxygen is absorbed by the blood; and that in the lung the oxygen merely enters into a state of loose combination with the blood, the union being intimate and complete only in the system. The complete chemical combination of the oxygen with the carbon takes place, then, not in the lungs, but in the capillary arteries of the system; consequently it is only while flowing in capillary arteries that carbonic acid is formed; that is, it is only in these vessels that the arterial combustion takes place: of course, therefore, it is only in these vessels that heat is extricated, and only from them that it can be communicated to the adjacent parts. According to this view, wherever there is a capillary artery, the combustion of carbon incessantly goes on, and there caloric is as incessantly set free; but since there is not a point of any tissue, in which there are not capillary arteries, there is not a point from which caloric does not radiate. As soon as formed, carbonic acid passes from the capillary arteries into the capillary veins; by the veins it is transmitted to the lungs; and by the lungs it is expelled from the system. The real operations car-

ried on in the lungs, then, are the transmission of oxygen and the extrication of carbonic acid; but this organ is not the seat of the essential and ultimate part of the function; it is merely the portal through which the elements employed in the process have their entrance and exit. Thus the question concerning the greater capacity of arterial blood for caloric is of no importance whatever: the phenomena may be equally accounted for, whatever be, in this respect, the constitution of the blood.

535. The result of the whole is, the complete establishment of the fact, that the production of heat in the animal body is a chemical operation, dependent on the combination of oxygen with carbon in the capillary arteries of the system; that is, it is the result of the burning of charcoal at every point of the body.

536. The agent which maintains and regulates this internal fire is the nervous system. There is, indeed, reason to suppose that the nervous system, in some mode or other, contributes to the actual production of animal heat. It is established by direct experiment, that the quantity of carbonic acid formed in the system is inadequate to the supply of the caloric expended by it; that in a given time more heat is abstracted from the body by the surrounding medium, than can be accounted for by the consumption of the amount of carbonic acid thrown off by the lungs during the same inter-

val. There is evidence that the source of this additional heat is the nervous system.

537. The influence exerted by the nervous system over the production of animal heat, is demonstrated by the fact, established by numerous observations and experiments, that whatever weakens the nervous power, proportionally diminishes the capacity of producing heat. For,

1. The destruction of a portion of the spinal cord diminishes the temperature of an animal without, as far as is ascertained, the disturbance of any other function.

2. The privation of the heart and blood-vessels of the nervous influence, as by decapitation, though the passage of the blood through the lungs and its ordinary change from the venous to the arterial state be maintained by artificial respiration, greatly diminishes, if it do not altogether suspend, the generation of animal heat.

3. The abolition of sensibility by the administration of a narcotic poison, artificial respiration being maintained, as effectually disturbs the generation of animal heat as decapitation; while the power of generating heat is restored, in the exact proportion to the return of the sensibility by the cessation of the action of the poison.

4. The temperature of an organ is found, by direct experiment, to be diminished by the division of the nerves that supply it with nervous

influence. The nerves that supply the horn were divided on one side of the body in a young deer; the other horn was left entire. The temperature of the horn—the nerves of which had been divided—was found, after some hours, to be considerably diminished, and it continued diminished for several days; at length its temperature was restored. On examining the horn about ten days after the operation had been performed, the divided nerves were found to be connected by a newly-formed substance; thus apparently accounting for the loss of temperature in the first instance, and for its subsequent restoration.

538. But although these and other analogous facts prove, beyond all question, the important influence of the nervous system over the development of animal heat, yet the mode in which that influence operates is not ascertained. Its action may be either direct or indirect. The nerves may possess some specific power of generating heat,—extricating it immediately from the blood by a process analogous to secretion,—or they may evolve it indirectly by other operations, as by some of the processes of nutrition. Each hypothesis is maintained by able physiologists; but the balance of evidence (as will appear hereafter) is greatly in favour of the opinion that the influence of the nervous system over this process is altogether indirect. A beautiful illustration of this is afforded in the following operation, which

is going on, without ceasing, every instant during life.

539. The skin which forms the external covering of the body is composed essentially of gelatin. No gelatin is contained in the blood; but the albumen of the blood is capable of being converted into gelatin by the addition of oxygen. Albumen is received by the capillary artery of the skin; the blood, of which albumen forms so important a constituent, contains a quantity of oxygen which it receives at the moment of inspiration, and which it retains in a state of loose combination (470 *et seq.*). Under the influence probably of the organic nerve, the capillary artery chemically combines a portion of the free oxygen with the albumen of the blood, and gelatin is the result. In this process the albumen gives off carbon; the blood affords oxygen; the two elements unite; carbonic acid is formed; and, as in every other instance in which carbonic acid is formed, heat is evolved. In this manner a fire is kindled, and is kept constantly burning, where it is most needed to counteract the influence of external cold, at the external surface of the body.

540. Such are the main points which have been established in relation to the production and distribution of animal heat. But it has been shown that the living body is capable of bearing without injury a temperature by which it is

rapidly consumed when deprived of life. By what means does the vital power enable the body to resist the influence of such intense degrees of heat?

541. Two circumstances are observable when the body is placed in a temperature greatly higher than its own. First, it can endure such a temperature only in the medium of air. Air can easily be borne at the temperature of 260°; aqueous vapour at the temperature of 130° few Europeans are capable of enduring longer than twelve minutes; the peasants of Finland appear to be able to sustain it, for the space of half an hour, as high as 167°; but the hottest liquid water-bath which any one seems to have been able to bear for the space of ten minutes, is the hottest spring at Barêges, the temperature of which is 113°. But in heated air the quantity of heat in actual contact with the body is much less than in the other media; because in proportion as the air is heated it is expanded, and in proportion as it is expanded the particles are diminished that come into contact with the body.

542. In the second place, the afflux of the colder fluids from the central parts of the system to the surface may for a time exert some influence in keeping down the temperature of the body. But above all this, in the third place, a two-fold provision is made in the body itself for the reduction of its temperature when exposed to in-

tense degrees of heat; by the one, the power with which it is endowed of producing heat is diminished; by the other, cold is positively generated.

543. It has been shown (517) that in proportion to the elevation of the temperature to which the body is exposed the blood becomes less venalized, and in the proportion in which the blood retains its arterial character the consumption of oxygen is diminished. Venous blood contains an excess of carbon, arterial blood an excess of oxygen. Consequently in proportion as the blood retains its arterial character it affords less carbon for the combination of oxygen, that is less inflammable matter. At an elevated temperature therefore there must, of necessity, be a diminished production of heat within the body, since the blood contains a diminished quantity of combustible material.

544. Moreover, in proportion to the elevation of the temperature to which the body is exposed, evaporation takes place from the entire surface of the pulmonary vesicles. No experiments have been performed which enable the physiologist to ascertain precisely the quantity of vapour exhaled from the lungs in a given time, when the body is exposed to a given degree of heat; but both observation and experiment show that it is very great. The blood pours out upon the whole surface of the air vesicles a

quantity of moisture in the form of water: by the surrounding air this water is converted into vapour: by the conversion of a fluid from the state of a liquid into that of vapour caloric is absorbed: by the absorption of caloric cold is generated, and that to such a degree that fluids exposed to the influence of evaporation may be frozen in the intensest heat of summer. The very process by which art, aided by science, affords to the inhabitants of warm climates the luxury of ice, is that by which nature generates cold in the human lungs when the body is exposed to a temperature above its own. Not only, then, is the lung the instrument by which the body acquires the power of evolving heat in greater or less quantity in proportion to the demands of the system, but this very same organ, under a change of circumstances, produces the directly contrary effect, and actually generates cold.

545. In the process of producing cold the skin is a powerful auxiliary to the lungs. More fluid is, indeed, evaporated from the surface of the skin in the form of perspiration, than from the lungs in the form of vapour; the cutaneous, like the pulmonary evaporation, increases in the ratio of the temperature, and both co-operate in abstracting the excess of caloric.

546. Finally, in proportion to the elevation of the temperature is the acceleration of the circulation; the pulse is augmented in power, and

doubled or trebled in frequency (495); but in proportion to the rapidity of the circulation is the increase of the quantity of evaporable matter which is transmitted to the evaporating surfaces.

547. From the whole it appears that by the combination of carbon and oxygen provision is made for the production of the greatest quantity of caloric that can at any time be required for the wants of the system; that when a decreased evolution of heat is necessary a smaller quantity of carbon and oxygen is brought into union, and that when, from exposure to intense degrees of heat, it is requisite for the maintenance of the temperature of the body at its own standard, that it should actually generate cold, it accomplishes this object by the evaporation of water.

CHAPTER X.

OF THE FUNCTION OF DIGESTION.

Process of Assimilation in the plant; in the animal—Digestive apparatus in the lower classes of animals; in the higher classes; in man—Digestive processes—Prehension, Mastication, Insalivation, Deglutition, Chymification, Chylification, Absorption, Fecation—Structure and action of the organs by which these operations are performed—Ultimate results—Powers by which those results are accomplished—Two kinds of digestion, a lower and a higher; the former preparatory to the latter.

548. Digestion is the function by which the aliment is converted into nutriment. No food can nourish until it be converted into a fluid analogous in chemical composition to that of the body by which it is assimilated. The conversion of the crude aliment into such a fluid is effected by a vital power peculiar to living beings, by which they subvert the constitution of other organized bodies, and cause them to assume their own. They accomplish this change by the agency of certain secretions which they elaborate in their own organs, and which they add to the substances they receive as aliment. By the action of these secretions, the chemical composition of the ali-

ment is brought into a close affinity to that of the body which it nourishes.

549. This change in the chemical composition of the aliment, by means of fluids secreted by the living bodies which receive it, is manifest in the plant as well as in the animal. The sap, as it issues from the root, is a colourless and limpid fluid; it has a specific gravity a little greater than that of water; it has a sweetish taste; it contains an acid which is sometimes free, and is either the carbonic or the acetic; but more commonly it is combined with lime or potass. To this crude sap, in this the first stage of its formation, vegetable secretions, sugar and mucus, assimilative substances, are superadded, probably by the fibres of the root.

550. As the sap ascends in the stalk, a greater quantity and a greater number of these vegetable secretions are poured into it. In the ratio of its elevation it acquires sugar, mucus, albumen, and an azotized substance analogous to gluten. By the admixture of these assimilative secretions, the crude sap is progressively assimilated nearer and nearer to the chemical composition of the proper nutritive fluid of the plant. Thus prepared, the sap passes to the leaf, in the upper surface of which it undergoes a process analogous to that of digestion in the animal (315), and is converted into proper nutrient matter.

551. The plant can only take up, by absorp-

tion, liquid food; it never receives solid substances as aliment: it therefore needs no apparatus for the division, solution, and fluidification of its food; its sole work of assimilation consists in changing the innate affinities of liquid aliment. But animals which live on vegetable and animal substances have to modify, by their digestive juices, the affinities of organic solids: hence assimilation in the animal must necessarily be a more complex operation than it is in the plant.

552. Fixed immovably to the soil by its roots, the nutritive apparatus of the plant is always in contact with its food, which is slowly but un-unceasingly absorbed according to the wants of its system. But the animal endowed with the faculty of locomotion receives its aliment into the interior of its body, that it may transport its food along with it in all its changes of place; and that, as in the plant, its food may be always in contact with its nutritive apparatus. The interior nutrition of the animal and the convergence of its nutritive apparatus to the centre of its system, and the exterior nutrition of the plant and the divergence of its nutritive apparatus to the peripheral extremity of its body, are differences in their mode of nutrition, connected with essential differences in the mode of life peculiar to the two beings.

553. Plant-like animals have a plant-like mode of nutrition. The transition from the one

class to the other is so gradual as to be almost insensible. Fixed to the same spot in the ocean as the tree to the land, the nutritive surface of the poriferous animal is always in contact with the water, as the soil is with the external surface of the plant. The cellular substance of which the bag of the poriferous animal is composed is permeated in all directions by ramifying and anastomosing canals, which, beginning by minute pores placed on the external surface, terminate in larger orifices, termed vents, which are fecal openings. These internal canals are incessantly traversed by streams of water, which enter through the minute, and are discharged through the larger orifices. By these currents the nutrient matter contained in the water is conveyed to every part of the body, and the streams that issue from the fecal orifices abound with minute flocculent particles, the residue of the digested matter. No separate part of the body is appropriated to the function of digestion any more than in the plant; there is merely a general absorbent surface; the water is to this animal what the soil is to the plant; its whole surface is a root; every point of that surface is constantly in contact with its food, and every point is absorbent.

554. In the class above the porifera, the margins of the superficial pores are merely lengthened out into minute sacs, irritable and sentient, surrounded with vibratile cilia (342). These sacs,

which are termed polypi, are so many little stomachs, which select, seize, and digest the food brought to them in the currents of water created by the action of the cilia (344).

Fig. CXLVIII.—*Hydra Viridis.*

1. The Hydra with its tentacula expanded. 2. The tentacula. 3. The body of the Hydra. 4. Disc for attachment. 5. The Hydra in the act of creeping. 6. The Hydra with an animalcule in its digestive cavity.

555. The fresh-water polype, the little hydra (fig. CXLVIII. 1), is one of these minute sacs detached and endowed with the power of locomotion (fig. CXLVIII. 5), a sentient, self-moving digestive bag. Capable of swallowing animals many times its own size, as the red-blooded worm, this little creature stretches its whole body like a thin elastic membrane over its prey, so as completely to alter its own shape, and the membranous sub-

stance of which it is composed becoming trans-parent by the distention, allows the subsequent process to be distinctly seen. The red fluid of the worm, as the process of digestion advances, is slowly diffused over every part of the internal surface of the polype. The whole internal surface of this minute self-moving bag is digestive; a true and proper stomach (fig. CXLVIII. 6). By dexterous manipulation, this internal surface may be rendered external, and the animal turned completely inside out. Then the external begins to perform the office of the internal surface, carrying on the function of digestion, just as well as that which was primitively formed for it; while the originally digestive becomes the generative surface, for the creature buds from this surface, now the outer one; a striking and instructive illustration of the analogy between the external covering of the animal body or the skin, and its internal lining, or the mucous surface.

556. In the monades (fig. CXLIX.), and in all

Fig. CXLIX.

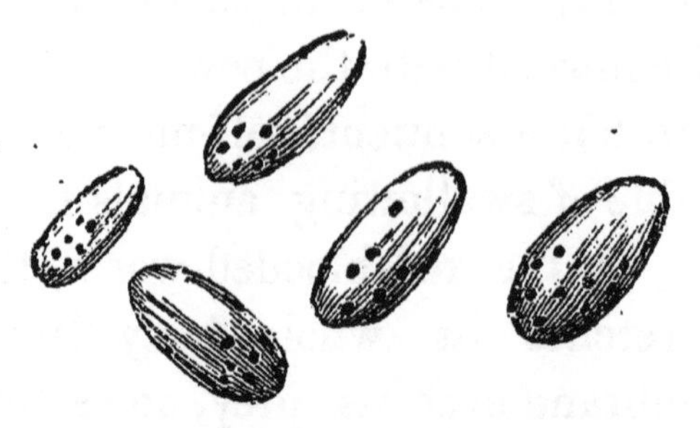

Group of Monades; the dark spots in the interior of their bodies representing their digestive sacs.

the lower animalcules, the digestive apparatus, instead of forming the entire internal surface of the body, consists of numerous sacs, which constitute so many separate stomachs, whence the name of the class, *polygastrica*. When empty, or when filled with water, these digestive sacs cannot be distinguished from the common cellular tissue of the body; but on feeding the animals with coloured organic matter, minutely diffused in water, the coloured particles readily enter the digestive sacs, and render apparent their form and arrangement. In the minutest animal hitherto appreciable, the monas termo, the 2000th part of a line in diameter, four rounded sacs have been seen filled with coloured particles (fig. CXLIX.). Each of these sacs, about the 6000th part of a line in diameter, opens by a narrow neck into a funnel-shaped mouth, surrounded with a single row of long vibratite cilia, by the action of which the floating organic particles are brought within the reach of the mouth. In general, even in this class, an alimentary canal traverses the whole extent of the body, into which all the different stomachs open. Sometimes numerous branches proceed from the main trunks of the alimentary canal, bearing the nutritive matter to the different parts of the body (fig. CL. 2). Often, in order to extend the digestive surface, the alimentary canal is produced, forming rounded enlargements called cœcal appendages, all of which act as so many

Fig. CL.—*Fasciola Hepatica.*

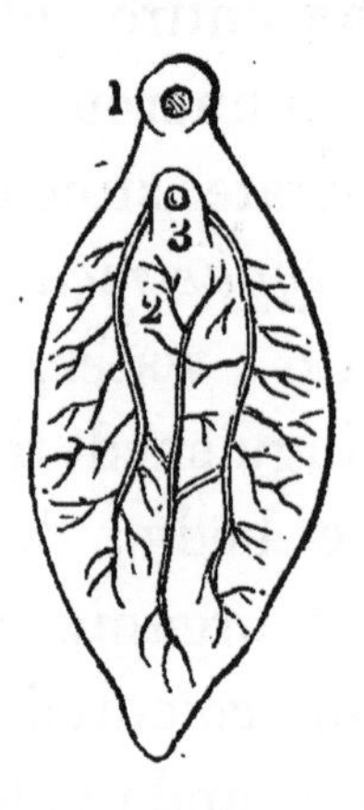

1. Mouth. 2. Alimentary tubes. 3. Sucker.

additional stomachs (fig. CLI. 3). In some individuals, observed under favourable circumstances, nearly 200 of these cœcal stomachs, filled with coloured matter, have been counted, and there may have been many more unseen, because empty and collapsed. In the lowest tribes of this class there is but one orifice to the alimentary canal, the oral; the food entering, and the fecal matter passing out of the system by the same aperture; but in the higher orders there is both an oral and an anal orifice, and the mouth and the anus are placed at opposite extremities of the body, as in the higher animals.

557. Up to this point in the animal series the digestive sacs and the alimentary canal are merely cavities formed in the common cellular tissue of the body, without any lining membrane, without teeth, or without any instruments for dividing

and preparing the aliment, and without a single gland, as far as has been ascertained, to assist the

Fig. CLI.—*Aphrodita Aculeata.*

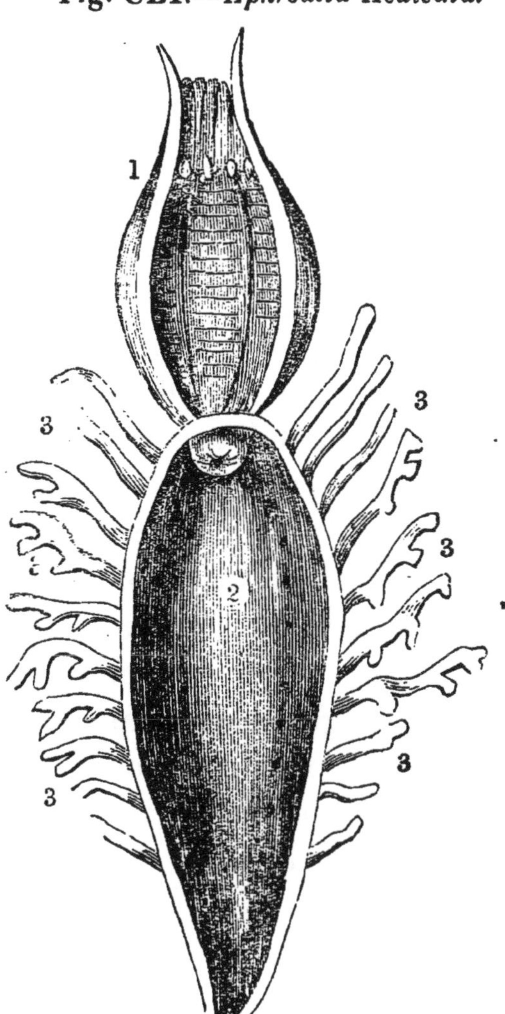

1. Proboscis in a retracted state. 2. Interior of digestive cavity. 3, 3. Cœcal appendages opening into it.

digestive process. All the assimilative functions, the respiratory as well as the digestive, appear to be performed by this single surface. But in the ascending scale not only is an apparatus appropriated to digestion, perfectly distinct from that assigned to respiration, but even the stomach and the alimentary canal are separate organs, distinguished from each other, both in structure and function. Still higher in the scale new organs are successively added, as the process becomes more complex and refined, in order to assist the main operations carried on in particular parts of the apparatus ; and as that apparatus approaches its highest degree of perfection, not only do the several parts of which it is composed increase in number and complexity, but each part becomes more and more isolated from the rest, a specific office being assigned to each in the division of labour that is made. Viewing, however, the digestive apparatus as a whole, whether simple or complex, whether consisting of a single uninterrupted surface, or divided into many separate portions, its nature is universally and invariably the same, and from the monad to man is endowed with analogous vital energies.

558. Comparative anatomy, which has succeeded in tracing through the different classes, orders, genera, and countless tribes of animals, the modifications in form and structure of the digestive apparatus, has shown that those modifications are invariably in strict adaptation to the kind of food on which the apparatus is destined to act

and to the extent of the elaboration requisite to convert crude aliment into proper animal substance. To trace this adaptation through the rising and ever-varying series, is a most interesting and instructive study, not only exhibiting, in the very organs that elaborate its food, the physical and even the mental qualities assigned by the hand of nature to each individual, but oftentimes shedding a clear and bright light on the complex structures of the highest and most perfect organization. Striking and beautiful illustrations are afforded by these investigations of the principle formerly insisted on (vol. i. chap. i. p. 28, 3), that the communication of the higher faculties exalts the apparatus even of the very lowest processes, that the latter may work in harmony with the former. In conformity with this principle, as the nobler endowments exalt the animal in the scale of organization, so even its very digestive apparatus becomes extended, isolated, complex and refined.

559. The highest and most perfect form of the digestive apparatus is that which is disposed in a series of chambers in free communication with each other. In these chambers the food undergoes a succession of changes, by which it is progressively assimilated to the nature of animal substance. This assimilation, however, is never effected by the sole agency of the chambers themselves; it is accomplished, to a great extent, by

the influence of special organs placed in the neighbourhood of the digestive chambers. In the lowest animal there is but one substance and one surface for every function; in the highest, even for the performance of the lowest function, there is the combination of many substances which are arranged in complex modes.

560. In man, the digestive chambers are five; the auxiliary organs are many.

The first of these chambers is the cavity called the mouth; the second is the bag termed the pharynx; the pharynx communicates by the esophagus with the third chamber, the stomach; the fourth chamber consists of the convoluted tubes named the small intestines, and the fifth consists of the larger tubes, denominated the large intestines. The assistant organs are, first, numerous appendages to the mouth, namely, the tongue, the teeth, the salivary glands, and the muscles that work the jaws; and, secondly, certain appendages to the small intestines, namely, the pancreas, the liver, the mesenteric glands, and the lacteal vessels.

561. By the mouth the food is softened and reduced to a pulp; by the tongue, materially aided by the soft palate, this pulp, when duly prepared, is transmitted to the pharynx; received by the pharynx, it is sent on to the esophagus; by the esophagus, it is conveyed to the stomach; in the stomach, it is converted into a peculiar substance

called chyme; the chyme, passing from the stomach into the first portion of the small intestines, is there converted into the substance called chyle; the chyle, carried slowly along the remaining portion of the small intestines, is successively absorbed by the lacteals; by the lacteals, it is conveyed through the mesenteric glands to the thoracic duct, and by the thoracic duct it is poured into the venous blood close to the heart. By the large intestines the refuse matter is conveyed out of the system.

562. The function of digestion consists, then, of the following processes:—

1. Prehension. 2. Mastication. 3. Insalivation. 4. Deglutition. 5. Chymification. 6. Chylification. 7. Absorption. 8. Fecation.

563. Prehension is the reception of the aliment; mastication is the mechanical comminution of it; insalivation is the admixture of it with certain juices poured into the mouth; deglutition is the transmission of it, when duly moistened and divided, into the stomach; chymification is the conversion of it into chyme; chylification is the conversion of the chyme into chyle; absorption is the assumption of the chyle by the lacteals and the transmission of it into the blood, and fecation is the separation and discharge of the refuse matter. Each part of this extended apparatus is modified in structure so as specially to fit it for

the performance of the office which is appropriated to it.

564. The mouth is not merely the opening between the two lips, but consists of an oval chamber, bounded above by the upper jaw and the palate; below by the tongue and the lower jaw; laterally by the cheeks; behind by the soft palate; and before by the lips.

565. The upper and lower jaw, the palate bones, and the teeth, constitute the hard or the bony parts of the mouth. The soft parts consist of the lips, the cheeks, the soft palate, the tongue, and the mucous membrane which lines the whole.

566. The lips and cheeks are composed principally of muscles, covered on the outside by the skin, and lined on the inside by the mucous membrane of the mouth. In the interspaces between the muscles is disposed a quantity of fat, which gives form to the face, facilitates the movements of the muscles, and protects the glands, blood-vessels, and nerves, with which all these organs are most abundantly supplied.

567. The roof of the mouth, called the palate, consists partly of bony and partly of membranous substance. The bony part of the palate forms an arch in the upper jaw, the position of which in the erect posture is horizontal: the membranous part of the palate consists of the mucous membrane of the mouth, which affords a covering to the bony part of the palate.

568. From the posterior part of the bony arch of the palate is suspended, transversely, a moveable partition, called the soft palate (fig. CLII. 1 and 2), which is composed of muscular fibres enclosed in the mucous membranes of the mouth. No less than ten distinct muscles enter into the compo-

Fig. CLII.—*View of the Mouth, showing particularly the Soft Palate, Tonsils, and Tongue.*

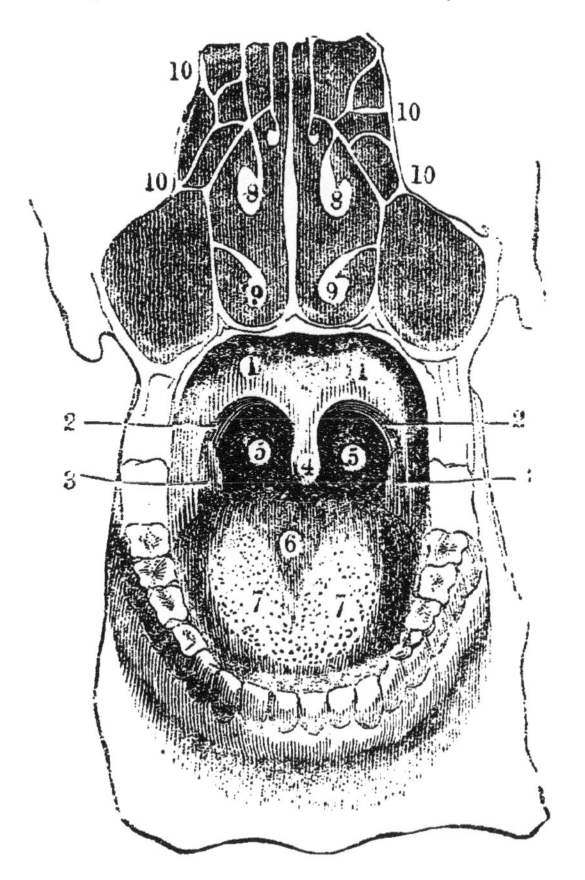

1. Anterior arch of the soft palate. 2. Posterior arch. 3. Tonsils or amygdalæ. 4. Uvula. 5. Communication between the mouth and pharynx. 6. The tongue. 7. Anterior or nervous papillæ. 8 and 9. The upper and lower turbinated bones dividing the nostrils into (10) chambers.

Fig. CLIII.—*A side view of the Mouth, Pharynx, Nose, &c.*

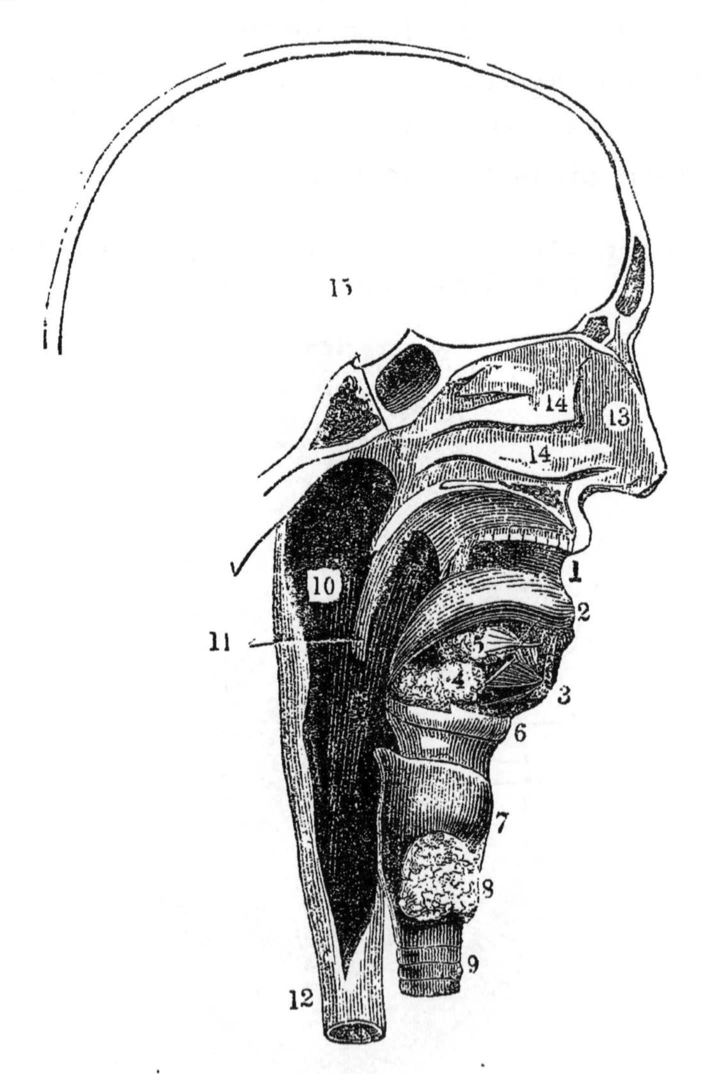

1. Mouth. 2. Tongue. 3. Section of the lower jaw. 4. Submaxillary gland. 5. Sublingual gland. 6. Hyoid bone. 7. Thyroid cartilage. 8. Thyroid gland. 9. Trachea. 10. Interior of the pharynx. 11. Section of the soft palate. 12. The esophagus. 13. The interior of the nose. 14. The two spongy bones dividing it into three chambers. 15. The posterior communication with the upper part of the pharynx.

Fig. CLIV.—*Posterior view of the Nose, Mouth, Larynx, and Pharynx laid open.*

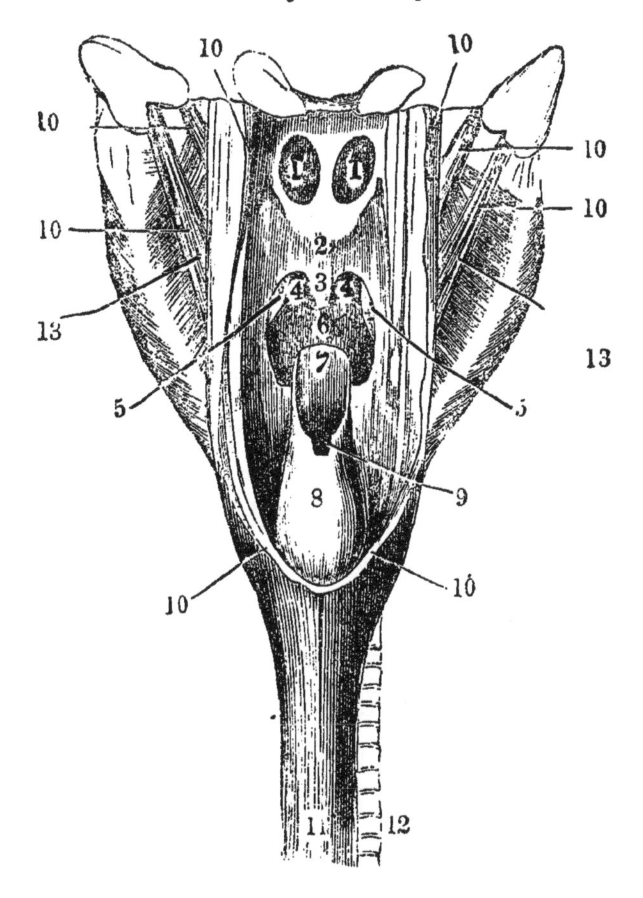

1. Posterior openings of the nose, communicating with the upper part of the pharynx. 2. Posterior surface of the soft palate. 3. The uvula. 4. Back part of the mouth communicating with the pharynx. 5. The tonsils. 6. Back part or root of the tongue. 7. Posterior surface of the epiglottis. 8. The larynx. 9. The opening of the larynx into the pharynx. 10. Cut edges of the pharynx. 11. Esophagus, the continuation of the pharynx. 12. The Trachea, continuation of the larynx. 13. Muscles acting on the pharynx.

sition of the soft palate. These muscles are disposed in such a manner that they render the organ capable of descending and of applying itself against the tongue (fig. CLII. 6), so as completely to close the passage between the mouth and the pharynx (figs. CLII. 5, and CLIV. 1), and of ascending and carrying itself obliquely backwards towards the posterior wall of the pharynx, so as completely to close the passage between the pharynx and the nose (fig. CLIV. 2, 1); hence this moveable partition performs the office of a double valve, closing the passage from the mouth to the pharynx, and from the pharynx to the nose.

569. From the centre of the soft palate hangs pendulous the conical-shaped body called the uvula (fig. CLII. 4), which consists of a small muscle enveloped in the mucous membrane of the mouth. The uvula assists in completing the valve formed by the soft palate (fig. CLIV. 2, 3); it is also an important organ in the modulation of the voice. When destroyed by disease, both the deglutition of the food and the sound of the voice become imperfect.

570. The lateral edges of the soft palate separate into two layers, which enclose between them the bodies called the tonsils (fig. CLII. 3), two glands commonly about the size of an almond. These organs co-operate with other glands in secreting the fluids of the mouth.

571. The tongue (figs. CLII. 6, and CLIII. 2) is composed of six distinct muscles enveloped in the mucous membrane of the mouth. The fibres

of these muscles are so interwoven with each other as to form an intricate net-work; and their number, arrangement, and exquisite organization render the organ capable of executing a surprising variety of movements with the most perfect precision, and with a velocity of which the mind can scarcely form a conception: some of these movements being requisite to bring the aliment under the operation of mastication, and others to produce articulate speech.

572. The tongue divided into base, apex, and dorsum, is supported by a bone called the hyoid bone (os hyoides) (figs. CXXXVI. 1, and CLIII. 6), which, unlike any other bone of the body, is placed at a distance from the general skeleton, and completely imbedded in muscles. This singularly posted and delicately constructed bone is not only connected with the tongue, but with many other highly important muscles, to which it affords a support and a lever.

573. Each jaw is provided with sixteen teeth (fig. CLV.), arranged with perfect uniformity, eight on each side of each jaw (fig. CLV.); those of the one side exactly corresponding with those of the other (fig. CLV.). The teeth, from the differences they present in their size, form, mode of connection with the jaw, and use, are divided into four classes, namely, on each side of each jaw, two incisors (figs. CLVI. and CLVII. 1, 2); one cuspid (figs. CLVI. and CLVII. 3); two bicuspid

(figs. CLVI. and CLVII. 4, 5); and three molar (figs. CLVI. and CLVII. 6, 7, 8).

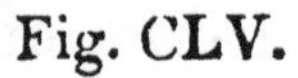

Fig. CLV.

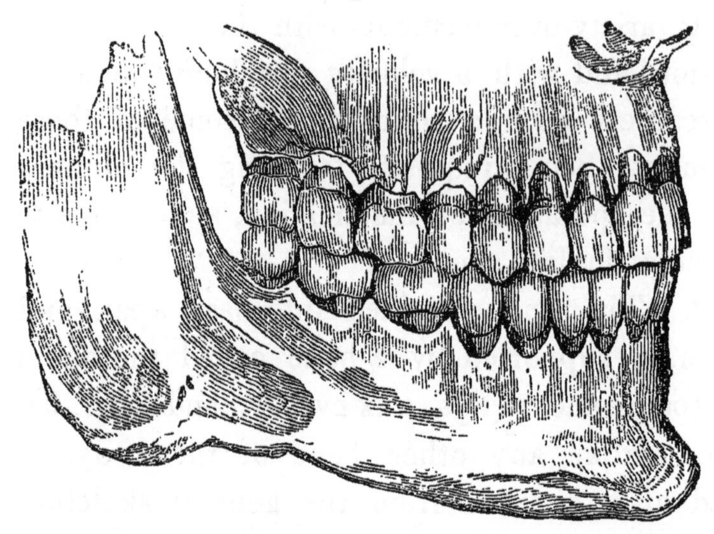

A lateral view of the whole series of the teeth, *in situ*, showing the relative situation of those of the upper with those of the lower jaw. This figure and the following figures to 159, are copied from Mr. T. Bell's scientific and instructive work on the Anatomy, Physiology, and Diseases of the Teeth.

574. The incisor, or cutting teeth, are situated in the front of the jaw; that directly in the centre is called the central; and the next to it the lateral incisor (fig. CLV.). Their office, as their name imports, is to cut the food, which they do, on the principle of shears or scissors.

575. Standing next to the lateral incisor is the cuspid, canine, or eye-tooth (figs. CLV., CLVI., and CLVII.). It is the longest of all the teeth. Its office is to tear such parts of the food as are too hard to be readily divided by the incisors.

576. Next the cuspid are the bicuspid, two on each side (fig. CLV., CLVII.), so named from their

Fig. CLVI.

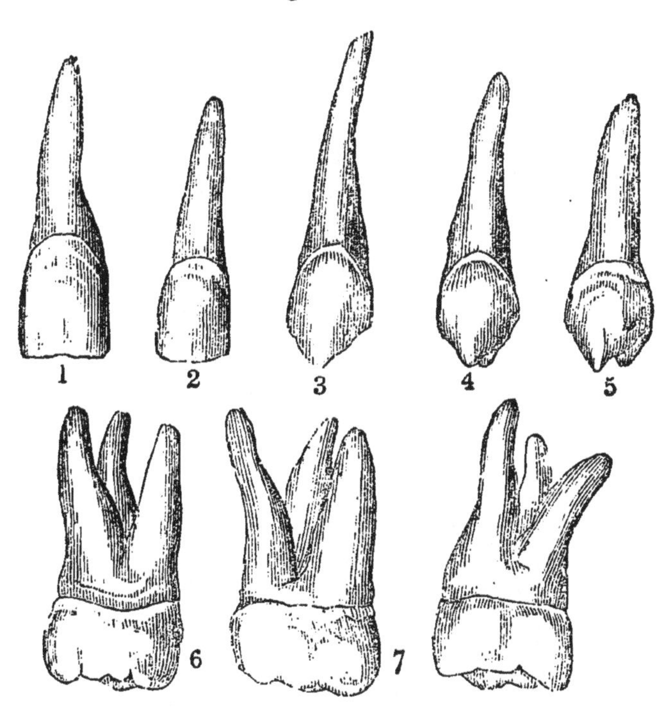

Front or external view of the upper teeth. 1. The central incisor. 2. The lateral incisor. 3. The cuspid. 4. The first bicuspid. 5. The second bicuspid. 6. The first molar. 7. The second molar. 8. The third molar, or dens sapientiæ.

being provided with two distinct prominences or points. Their office is to tear tough substances preparatory to their trituration by the next set.

577. The molars, or the grinders, three on each side (fig. CLVI. and CLVII.), provided with four or five prominences on the grinding surface, with corresponding depressions, which are so

arranged that the elevations of those of the upper are adapted to the concavities of those of the lower jaw, and the contrary.

578. From the incisor to the molar teeth there

Fig. CLVII.

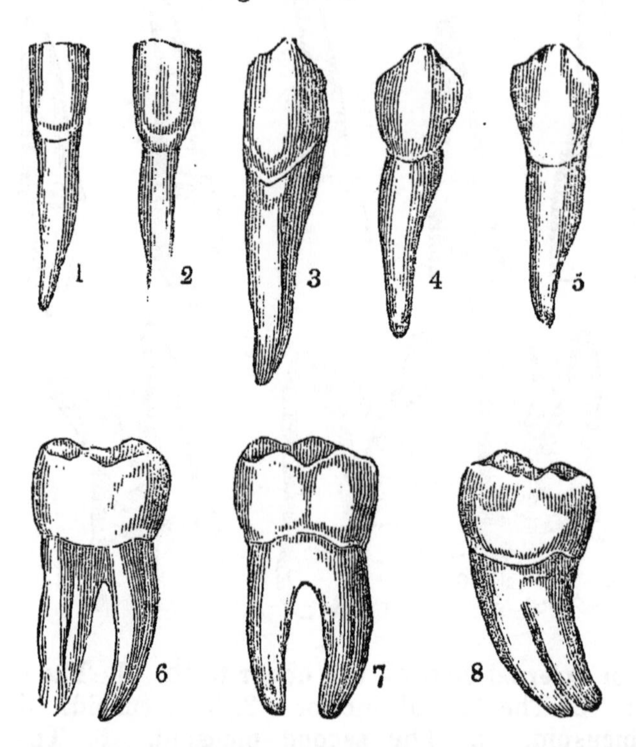

Front view of the lower teeth. 1. The central incisor. 2. The lateral incisor. 3. The cuspid. 4. The first bicuspid. 5. The second bicuspid. 6. The first molar. 7. The second molar. 8. The third molar, or dens sapientiæ.

is a regular gradation in size, form, and use, the cuspid holding a middle place between the incisor and the bicuspid, and the bicuspid being in every respect intermediate between the cuspid and the molar. Thus the incisor are adapted only for cutting, the cuspid for tearing, the bicuspid partly for tearing and partly for grinding, and the molar solely for grinding. The incisor has only a single root, which is nearly round, and quite simple (fig. CLVII. 1, 2); the cuspid has only a single root, but this is flattened and partially grooved (fig. CLVII. 3); even the bicuspid has only a single root, but this is commonly divided at its extremity, and is always so much grooved as to have the appearance of two fangs partially united, the body having two points instead of one, thus approaching it to the form of the molar (fig. CLVII. 4, 5); and these last have always two, sometimes three, occasionally four roots. and their body is greatly increased in size, and has a complete grinding surface (fig. CLVII. 6, 7, 8).

579. In some animals whose food and habits require the utmost extension of the office of a particular class of teeth, a corresponding development of that class takes place. Thus in the carnivora, as is strikingly seen in the tiger and the polar bear, the cuspid or canine teeth are prodigiously elongated and strengthened, in order to enable them to seize their food, and to tear it in pieces. On the other hand, in the rodentia, or gnawing animals, as in the beaver, the incisors are exceedingly elongated; while in the graminivora, and especially in the ruminantia, the molar teeth are by far the most developed. In each case the other kinds of teeth are of little comparative im-

portance; sometimes they are even altogether wanting. Thus the shark has only one kind of tooth, the incisor; but of these there are several rows, and all of them the creature has the power of erecting at will.

580. So intimately are these organs connected with the kind of food by which life is sustained, and the kind of food with the general habits of the animal, that an anatomist can tell the structure of the digestive organs, the kind of nervous system, the physical and even the mental endowments; that is, the exact point in the scale of organization to which the animal belongs, merely by the inspection of the teeth.

581. In man, the several classes of the teeth are so similarly developed, so perfectly equalized, and so identically constructed, that they may be considered as the true type from which all the other forms are deviations.

582. For the accomplishment of their office the teeth must be endowed with prodigious strength: for the fulfilment of purposes immediately connected with the apparatus of digestion, it is necessary that they should be placed in the neighbourhood of exceedingly soft, delicate, irritable, and sentient organs. That they may possess the requisite degree of strength, they are constructed chiefly of bone, the hardest organized substance. Bone, though not as sensible as some other parts of the body, is nevertheless sentient. The em

ployment of a sensitive body in the office of breaking down the hard substances used as food would be to change the act of eating from a pleasurable into a painful operation. It has been shown (vol. i. p. 84) that provision is made for supplying to the animal a never-failing source of enjoyment in the annexation of pleasurable sensations with the act of eating, and that, taking the whole of life into account, the sum of enjoyment secured by this provision is incalculable. But all this enjoyment might have been lost, might even have been changed into positive pain, nay, must have been changed into pain, but for adjustments numerous, minute, delicate, and, at first view, incompatible.

583. Had a highly-organized and sensitive body been made the instrument of cutting, tearing, and breaking down the food, every tooth, every time it comes in contact with the food, would produce the exquisite pain now occasionally experienced when a tooth is inflamed. Yet a body wholly inorganic and therefore insensible, could not perform the office of the instrument; first, because a dead body cannot be placed in contact with living parts without producing irritation, disease, and consequently pain; and, secondly, because such a body being incapable of any process of nutrition, must speedily be worn away by friction, and there could be no possibility of repairing or of replacing it. The instrument in question,

then, must possess hardness, durability, and, to a certain extent, insensibility; yet it must be capable of forming an intimate union with sentient and vital organs, must be capable of becoming a constituent part of the living system.

584. To communicate to it the requisite degree of hardness, the hard substance forming its basis is rendered so much harder than common bone that some physiologists have even doubted whether it be bone, whether it really possess a true organic structure. That there is no ground for such doubt the evidence is complete. For,

1. The tooth, like bone in general, is composed partly of an earthy and partly of an animal substance; the earthy part being completely removable by maceration in an acid, and the animal portion by incineration, the tooth under each process retaining exactly its original form.

2. The root of the tooth is covered externally by periosteum; its internal cavity is lined by a vascular and nervous membrane, and both structures are intimately connected with the substance of the tooth. If these membranes really distribute their blood-vessels and nerves to the substance of the tooth, which there is no reason to doubt, the analogy is identical between the structure of the teeth and that of bone.

3. Though the blood-vessels of the teeth are so minute that they do not, under ordinary circumstances, admit the red particles of the blood, and

though no colouring matter hitherto employed in artificial injections has been able, on account of its grossness, to penetrate the dental vessels, yet disease sometimes accomplishes what art is incapable of effecting. In jaundice the bony substance of the teeth is occasionally tinged with a bright yellow colour; and in persons who have perished by a violent death, in whom the circulation has been suddenly arrested, it is of a deep red colour. Moreover, when the dentist files a tooth, no pain is produced until the file reaches the bony substance; but the instant it begins to act upon this part of the tooth, the sensation becomes sufficiently acute.

585. These facts demonstrate that the bony matter of the tooth, though modified to fit the instrument for its office, is still a true and proper organized substance.

586. Each tooth is divided into body, neck, and root (fig. CLVIII. 1, 2, 3). The body is that part of the tooth which is above the gum, the root that part which is below the gum, and the neck that part where the body and the root unite (fig. CLVIII). The body, the essential part, is the tooth properly so called, the part which performs the whole work for which the instrument is constructed, to the production and support of which all the other parts are subservient.

587. When a vertical section is made in the tooth, it is found to contain a cavity of consider-

Fig. CLVIII.

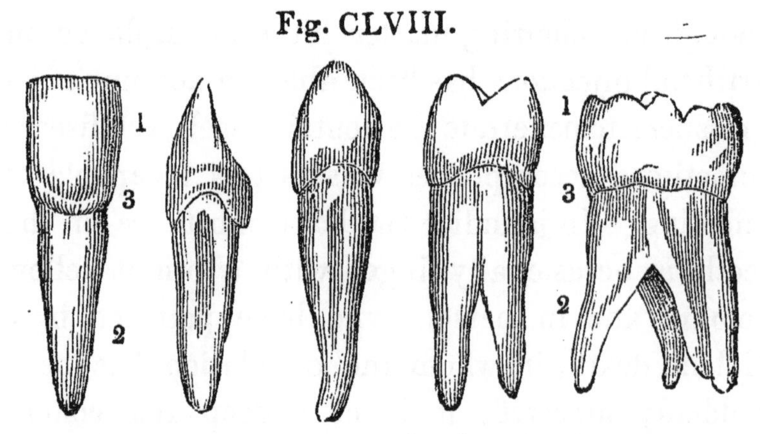

Views of different kinds of teeth, showing their anatomical division into, 1. The body or crown. 2. The fang or root. 3. The neck.

able size (fig. CLIX. 3), termed the dental cavity, which, large in the body of the tooth, gradually diminishes through the whole length of the root

Fig. CLIX.—*Sections of Teeth, exhibiting their Structure.*

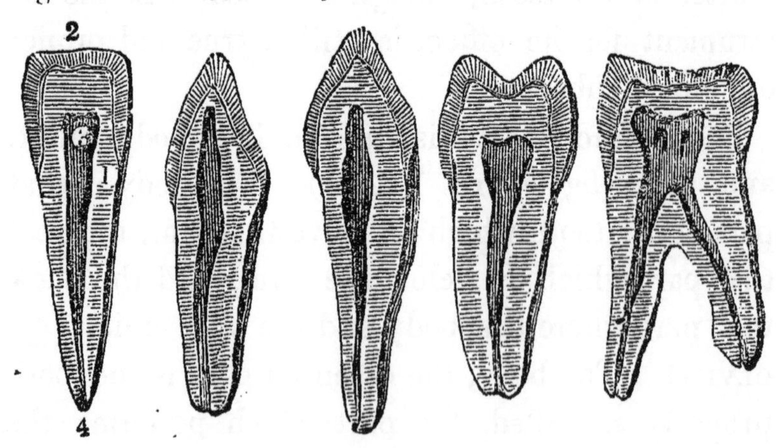

1. The bony substance. 2. The enamel. 3. The internal cavity. 4. The foramen, or hole at the extremity of the root.

(fig. CLIX. 3). The dental cavity is lined throughout with a thin, delicate, and vascular membrane, continued from that which lines the jaw. It contains a pulpy substance. This pulp, highly vasculai and exquisitely sensible, is composed almost entirely of blood-vessels and nerves, and is the source whence the bony part of the tooth derives its vitality, sensibility, and nutriment. The blood-vessels and nerves that compose the pulp enter the dental cavity through a minute hole at the extremity of the root (fig. CLIX. 4). The membrane which lines the dental cavity is likewise continued over the external surface of the root, so as to afford it a complete envelope.

588. Provision having been thus made for the organization of the tooth, for the support of its vitality, and for its connexion with the living system, over all that portion of it which is above the gum, and which constitutes the essential part of the instrument, there is poured a dense, hard, inorganic, insensible, all but indestructible substance, termed enamel (fig. CLIX. 2); a substance inorganic, composed of earthy salts, principally phosphate of lime with a slight trace of animal matter: a substance of exceeding density, of a milky-white colour, semi-transparent, and consisting of minute fibrous crystals. The manner in which this inorganic matter is arranged about the body of the tooth is worthy of notice. The crystals are disposed in radii springing from the centre of the tooth (fig. CLX. 3); so that the extremities of the crystals form the external surface of the tooth,

while the internal extremities are in contact with the bony substance (fig. CLX. 3). By this arrangement a two-fold advantage is obtained; the enamel

Fig. CLX.

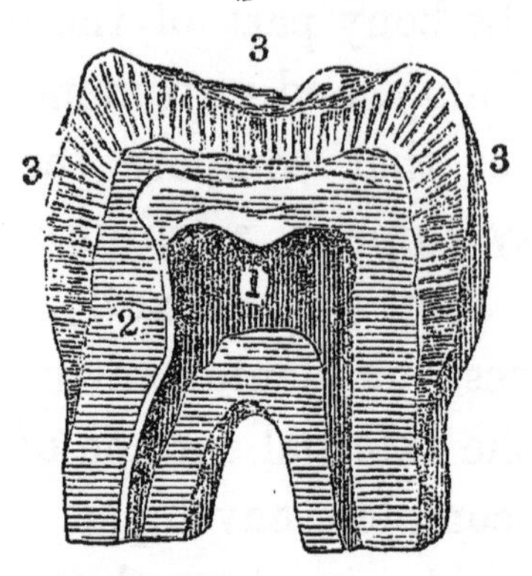

Magnified section of a tooth, to illustrate the arrangement of the fibrous crystals composing the enamel. 1 Cavity of the tooth. 2. Bony substance. 3. Enamel, showing the crystals disposed in radii.

is less apt to be worn down by friction, and is less liable to accidental fracture.

589. In this manner an instrument is constructed possessing the requisite hardness, durability, and insensibility; yet organized, alive, as truly an integrant portion of the living system as the eye or the heart.

590. No less care is indicated in fixing than in constructing the instrument. It is held in its situation not by one expedient, but by many.

1. All along the margin of both jaws is placed a bony arch, pierced with holes, which constitute the sockets, called alveoli, for the teeth (fig. CLXI.). Each socket or alveolus is distinct,

there being one alveolus for each tooth (fig. CLXI.). The adaptation of the root to the alveolus is so exact, and the adhesion so close, that each root is fixed in its alveolus just as a nail is fixed when driven into a board.

Fig. CLXI.

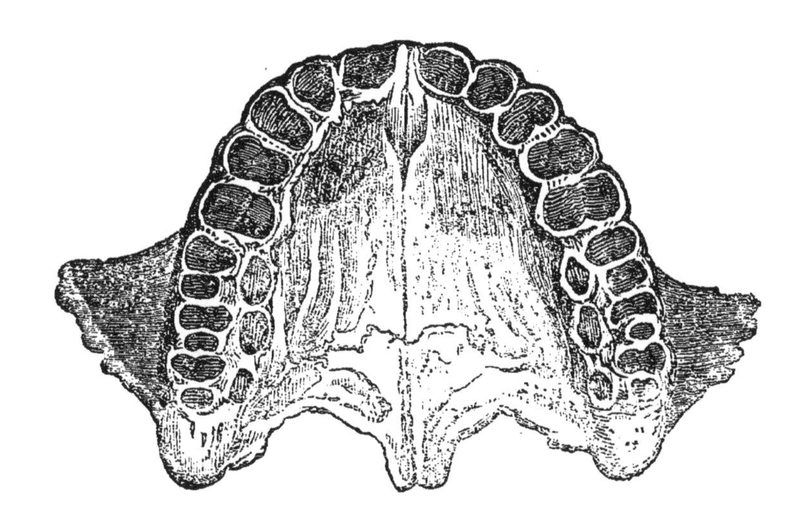

Upper jaw, showing the alveoli.

2. The roots of the tooth, when there are more than one, deviate from a straight line (fig. CLVI. 6, 7, 8) ; and this deviation from parallelism, on an obvious mechanical principle, adds to the firmness of the connexion.

3. Adherent by one edge to the bony arch of the jaw, and by the other to the neck of the tooth, is a peculiar substance, dense, firm, membranous, called the gum, less hard than cartilage, but much

harder than skin, or common membrane; abounding with blood-vessels, yet but little sensible; constructed for the express purpose of assisting to fix the teeth in their situation.

4. The dense and firm membrane covering the bony arch of the jaw is continued into each alveolus which it lines; from the bottom of the alveolus this membrane is reflected over the root of the tooth, which it completely invests as far as the neck, where it terminates, and where the enamel begins: this membrane, like a tense and strong band, powerfully assists in fixing the tooth.

5. Lastly, the vessels and nerves which enter at the extremity of the root, like so many strings, assist in tying it down; hence, when in the progress of age, all the other fastenings are removed, these strings hold the teeth so firmly to the bottom of the socket, that their removal always requires considerable force.

591. But a dense substance like enamel, acting with force against so hard a substance as bone, would produce a jar which, propagated along the bones of the face and skull to the brain, would severely injure that tender organ, and effectually interfere with the comfort of eating.

592. This evil is guarded against,

1. By the structure of the alveoli (fig. CLXII.), which are composed not of dense and compact, but of loose and spongy bone (fig. CLXII.). This can-

Fig. CLXII.

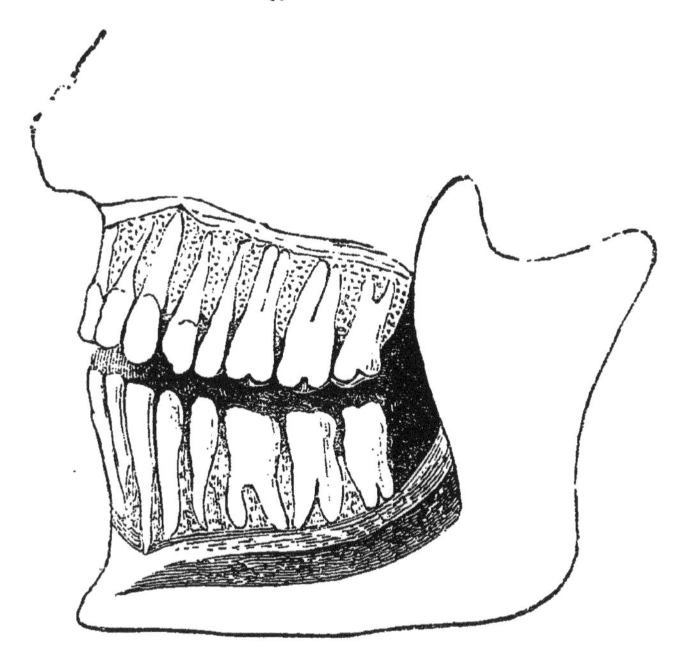

View of the upper and lower teeth in the alveoli; the external alveolar plate being cut away to show the cancellated structure of the alveoli, and the articulation of the teeth.

cellated arrangement of the osseous fibres is admirably adapted for absorbing vibrations and preventing their propagation (90).

2. By the membrane which lines the socket.

3. By the membrane which covers the root of the tooth; and,

4. By the gum.

These membranous substances, even more than the cancellated structure of the alveoli, absorb vibrations and counteract the communication of a shock to the bones of the face and head when the teeth act forcibly on hard materials; so many and such nice adjustments go to secure enjoyment, nay to prevent exquisite pain, in the simple operation of bringing the teeth into contact in the act of eating.

593. The teeth in mastication are passive instruments put in motion by the jaws. The upper jaw is fixed, the lower only is movable. The lower jaw is capable of four different motions; depression, elevation, a motion forwards and backwards, and partial rotation. These simple motions

Fig. CLXIII.—*View of the Muscles of Mastication, which elevate the lower jaw.*

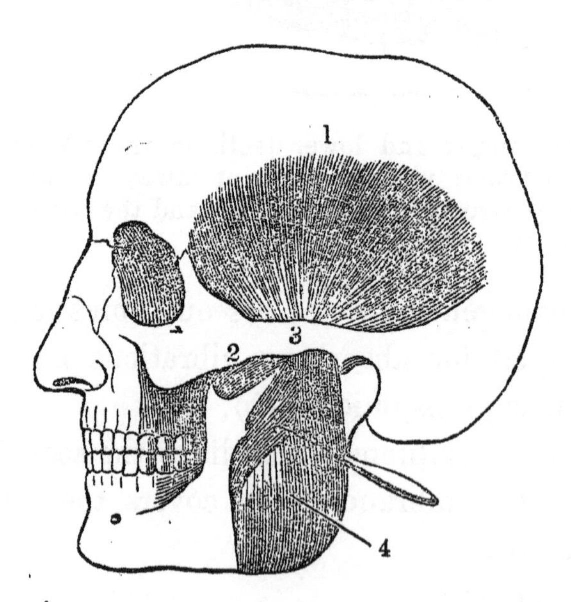

1. The temporal muscle. 2. Its insertion passing beneath. 3. The zygoma. 4. The masseter muscle, its anterior portion reflected to show the insertion of the temporal. The action of these powerful muscles is to pull the lower jaw upwards with great force against the upper jaw, and at the same time to draw it a little forwards or backwards, according to the direction of the fibres of the muscles.

are capable, by combination, of producing various compound motions. Numerous muscles, some of them endowed with prodigious power, are so disposed and combined as to be able, at the command

Fig. CLXIV.—*Muscles of the Jaw.*

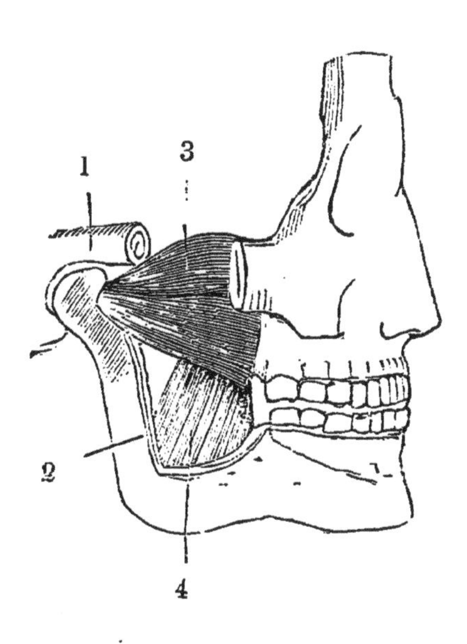

1. Portion of the zygomatic process of the temporal bone. 2. Ascending plate of the lower jaw removed to expose, 3. External pterygoid, and, 4. Internal pterygoid muscles. The action of these muscles is to raise the lower jaw, and to pull it obliquely towards the opposite side. When both muscles act together, they bring the lower jaw forwards, so as to make the fore-teeth project beyond those of the upper jaw.

of volition, to produce any of these motions that may be required, simple or compound.

594. By the combination, succession, alternation, and repetition of these motions, the lower is made to produce upon the upper jaw all the

variety of pressure necessary for the mastication of the food. In this process the muscles of the tongue perform scarcely a less important part than the muscles of the lower jaw. Some of its muscular fibres shorten the tongue, some give it breadth, others render it concave, and others convex: so ample is the provision for moving this organ to different parts of the mouth and fauces, whether to bruise the softer parts of the aliment against the palate, to mix it with the saliva, or to place it under the pressure of the teeth.

595. By the combined action of the muscles of the lower jaw and tongue, and that of the teeth, the food is bruised, cut, torn, and divided into minute fragments. This operation is of so much importance that the whole process of digestion is imperfect without it. It is proved by direct experiment that the stomach acts upon the aliment with a facility in some degree proportionate to the perfection with which it is masticated. If an animal swallow morsels of food of different bulks, and the stomach be examined after a given time, digestion is found to be the most advanced in the smallest pieces, which are often completely softened, while the larger are scarcely acted upon at all.

596. At the same time that, by the operation of mastication, the aliment undergoes mechanical division, it imbibes a quantity of fluid derived from various sources.

1. From the membrane which lines the internal surface of the mouth, and which affords a covering to all the parts contained in it.

2. From numerous minute glands placed in clusters about the cheeks, gums, lips, palate, and tongue. Each of these glands is furnished with its own little duct, which, piercing the mucous membrane, opens into the cavity of the mouth. From these glands is derived the fluid with which the interior of the mouth is lubricated. It consists of a glutinous, adhesive, transparent fluid, of a light grey tint, salt taste, and slightly alkaline nature, termed mucus.

Fig. CLXV.—*View of the Parotid Gland with the Muscles of the Face.*

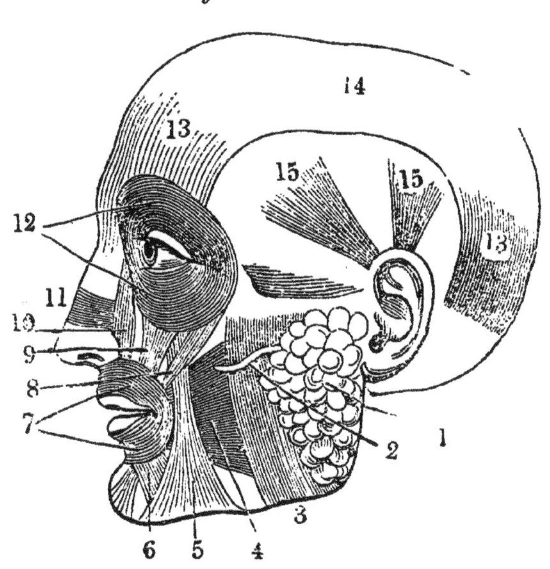

1. Parotid gland. 2. Parotid duct. 3. Masseter muscle. 4. Buccinator. 5. Triangularis, or depressor of the angle of the mouth. 6 Depressor of the lower lip. 7. Orbicularis, or circular muscle of the mouth. 8. Great zygomatic,

or the distorter of the mouth, as in laughing. 9. Elevator of the angle of the mouth. 10. Elevator of the upper lip, and wing of the nose. 11. Compressor of the cartilage of the nose. 12. Orbicularis, or circular muscle of the eye-lids. 13. Occipito frontalis; elevator of the eye-lids; motor of the scalp, &c., an important muscle of expression. 14. Tendinous portion of tho occipito frontalis. 15. Elevator of the ear.

3. From six large glands placed symmetrically, three on each side, termed the salivary glands, namely, the parotid (fig. CLXV. 1), situated before the ear; the submaxillary (fig. CLIII. 4), situated beneath the lower jaw; and the sublingual (fig. CLIII. 5), situated immediately under the tongue. Each of these glands is provided with a duct (figs. CLXV. 2, and CLIII. 4, 5), by which it pours the fluid it elaborates, called saliva, into the mouth.

597. The other fluids of the mouth are always mixed with the saliva, and are all commonly included under that name. The secretion of these fluids is unceasing, and they pass into the stomach by successive acts of deglutition at nearly regular intervals; so that the stomach, after it has been some time without food, contains a considerable quantity of these fluids. But they are chiefly needed during the operation of mastication, and two provisions are made for securing their flow in the greatest abundance at that time.

598. First, the situation of the glands is such that they are all exposed to the action of the muscles of mastication (figs. CLXIII. and CLXIV.),

by which action the glands are excited, a large quantity of blood is determined to them, and the quantity of fluid they secrete is proportionate to the quantity of blood they receive. Secondly, the glands are placed under the influence of the mind, so that the very thought, and still more the taste, of grateful food, acting upon them as an additional stimulus, causes an additional secretion. The quantity of fluid formed from these different sources, and mixed with the food during the mastication of an ordinary meal, is estimated at half a pint. It must commonly be more than this, because, in a case described by Dr. Gairdner, of Edinburgh, in which the esophagus had been cut through, it was observed that from six to eight ounces of saliva were discharged during a meal, which consisted merely of broth injected through the divided esophagus into the stomach.

599. Saliva is a frothy, watery fluid, in its healthy state nearly insipid, and of a slightly alkaline nature. It is composed of water, a peculiar animal substance called salivary matter, mucus, osmazome, a little albumen, and several salts. It produces important changes on the food. By the water, and the salts contained in it, it softens and dissolves the food; and thus, while it renders it easier to be swallowed, it prepares it for the subsequent changes it is to undergo. To this latter object, the assimilation of the food, it seems to communicate the first tendency by the azotized

substances, the salivary, and the albuminous matter which it adds to it. From this, the commencement of the assimilative process to its completion, animalized substances are successively added to the food which have the property of converting the food more and more into the nature of animal substance.

600. Comminuted by the teeth, and softened by the saliva, the food is reduced to a pulp. In this pulp there is a complete admixture of all the alimentary substances with the assimilative matter secreted from the blood, into the nature of which it is to be ultimately changed. The mass is at the same time brought to the temperature of the blood.

601. As long as the operations of mastication and insalivation go on, the mouth forms a closed cavity from which the food cannot escape; for the lips enclose it before, the cheeks at the sides, the tongue below, and the soft palate behind, the inferior edge of which being applied in close and firm contact with the base of the tongue, prevents all communication between the mouth and the pharynx.

602. When, by mastication, the food is sufficiently divided, and by insalivation softened and animalized to fit it for the future changes it is to undergo, it is collected by the tongue, and carried by that organ to the back part of the mouth. The soft palate (fig. CLII. 1), obedient to the stimulus

of the duly prepared food, rises the instant it is touched by it, and affords it a free passage to the pharynx (figs. CLIII. 10, and CLIV. 10).

603. The pharynx (fig. CLIII. 10), a muscular bag, immediately continuous with the mouth (fig. CLIII. 1), is a vestibule into which open several highly important organs. Before is the entrance to the windpipe, termed the glottis (fig. CLIV. 9), leading directly to the larynx (fig. CLIV. 8); at the sides are the mouths of two ducts, termed the Eustachian tubes, which lead to the internal part of the organ of hearing; above are two entrances to the nose (fig. CLIV. 1), and below is the passage to the stomach (fig. CLIII. 12).

604. Were the food to enter the Eustachian tubes or the nose, it would occasion great inconvenience; were it to enter the glottis, it would cause death. It is prevented from entering the Eustachian tubes and the nose by the soft palate (fig. CLII. 1 and 2), which by the very act of rising to afford an opening from the mouth to the pharynx, is carried over the other apertures so as completely to close them. By the varied direction of the muscular fibres which enter into the composition of this organ, it is enabled to execute the different and even opposite motions required in the performance of its important office.

605. The food is prevented from entering the glottis partly by a cartilaginous valve (fig. CLIV. 7), termed the epiglottis, placed immediately above the glottis, and attached to the root of the tongue

(fig. CLIV. 6). In delivering the food to the pharynx the tongue passes backwards (fig. CLIV. 6). In passing backwards it pushes in the same direction the epiglottis which is attached to it, and so necessarily carries it over the glottis, completely closing the aperture (fig. CLIV. 9). At the same time the opening is still more securely closed by the glottis itself, in consequence of the powerful and simultaneous contraction of the muscles that act upon it in the production of the voice. It is proved, by direct experiment, that the spontaneous closure of the glottis is a more powerful agent in excluding the food from the larynx even than the depression of the epiglottis; but both organs concur in producing the same result; and a double security is provided against an event which would be fatal.

606. It is deeply interesting to observe the part performed in these operations by sensation and volition, and the boundary at which their influence terminates and consciousness itself is lost. Mastication, a voluntary operation, carried on by voluntary muscles, at the command of the will, is attended with consciousness, always in the state of health of a pleasurable nature. To communicate this consciousness, the tongue, the palate, the lips, the cheeks, the soft palate, and even the pharynx, are supplied with a prodigious number of sentient nerves. The tongue especially, one of the most active agents in the operation, is supplied with no

less than six nerves derived from three different sources. These nerves, spread out upon this organ, give to its upper surface a complete covering, and some of them terminate in sentient extremities visible to the naked eye. These sentient extremities, with which every point of the upper surface, but more especially the apex, is studded, constitute the bodies termed papillæ, the immediate and special seat of the sense of taste. This sense is also diffused, though in a less exquisite degree, over the whole internal surface of the mouth. Close to the sense of taste is placed the seat of the kindred sense of smell. The business of both these senses is with the qualities of the food. Mastication at once brings out the qualities of the food and puts the food in contact with the organs that are to take cognizance of it. Mastication, a rough operation, capable of being accomplished only by powerful instruments which act with force, is carried on in the very same spot with sensation, an exquisitely delicate operation, having its seat in soft and tender structures, with which the appropriate objects are brought into contact only with the gentlest impulse. The agents of the coarse and the delicate, the forcible and the gentle operations are in close contact, yet they work together not only without obstruction, but with the most perfect subserviency and co-operation.

607. The movements of mastication are produced, and, until they have accomplished the

objects of the operation, are repeated by successive acts of volition. To induce these acts, grateful sensations are excited by the contact of the food with the sentient nerves so liberally distributed over almost the whole of the apparatus. To the provision thus made for the production of pleasurable sensation, is superadded the necessity of direct and constant attention to the pleasure included in the gratification of the taste. It is justly observed by Dr. A. Combe, that without some degree of attention to the process of eating, and some distinct perception of its gratefulness, the food cannot be duly digested. When the mind is so absorbed as to be wholly unconscious of it, or even indifferent to it, the food is swallowed without mastication; then it lies in the stomach for hours together without being acted upon by the gastric juice, and if this be done often, the stomach becomes so much disordered as to lose its power of digestion, and death is the inevitable result: so that not only is pleasurable sensation annexed to the reception of food, but the direct and continuous consciousness of that pleasurable sensation during the act of eating is made one of the conditions of the due performance of the digestive function.

608. With the operation of mastication and one part of the process of deglutition, immediately to be noticed, the agency of volition and sensation cease. Beyond this the function of digestion is

wholly an organic process. In addition to the reasons assigned (vol. i. p. 55) why all the organic processes are placed alike beyond the cognizance of sense and the control of the will, there is this special reason why, in the function of digestion, they cease at the exact boundary assigned them.

609. Every time the act of deglutition is performed the openings to the windpipe and to the nose are closed, so that during this operation all access of air to the lungs is stopped, consequently it is necessary that the passage of the food through the pharynx should be rapid. Mastication, a voluntary process, may be performed slowly or rapidly, perfectly or imperfectly, without serious mischief; but life depends on the passage of the food through the pharynx with extreme rapidity and with the nicest precision. It is therefore taken out of the province of volition and entrusted to organs which belong to the organic life, organs which carry on their operations with the steadiness, constancy, and exactness of bodies whose motions are determined by a physical law.

610. No sooner does the duly-prepared food touch the soft palate than the whole apparatus of deglutition is instantly in motion. This movable partition suddenly rises to afford to the food a free passage to the pharynx. The pharynx itself, at the same instant, rises to receive the morsel thrust towards it by the pressure of the tongue; and one muscle, the stylo-pharyngeus, which concurs in

producing this movement, seems specially intended, in addition, to expand the pharynx. Three muscles throw their fibres around the pharynx, termed its upper, middle, and lower constrictors, which, the moment the morsel reaches the pharynx, contract upon it, and embrace it firmly. At the same instant the larynx, closing its aperture, springs forward towards the base of the tongue, under which it is in a manner concealed, the additional shield of the epiglottis being simultaneously thrown over the glottis. By this movement of the larynx, upwards and forwards, the course of the morsel across the dangerous passage is shortened. All these motions take place with such rapidity that Boerhaave said the action is convulsive. And now the food, firmly pressed by the pharynx, cannot return to the mouth, for the root of the tongue is there stopping up the passage; it cannot enter the Eustachian tubes or the nose, for the soft palate is there closing the apertures; it cannot enter the larynx, for a double guard is placed upon the glottis securing its firm closure. The food can advance in one direction only, the direction required, that which leads to the esophagus. Well, therefore, on the contemplation of these complex structures and the consent and harmony with which they act, might Paley say, "In no apparatus put together by art do I know such multifarious uses so aptly contrived as in the natural organization of the human mouth and its appendages. In this small cavity we have teeth of different shape; first, for cutting; secondly, for grinding; muscles most artificially disposed for carrying on the compound motions of the lower jaw by which the mill is worked; fountains of saliva springing up in different parts of the cavity for the moistening of the food while the mastication is going on; glands to feed the fountains; a muscular contrivance in the back part of the cavity for the guiding of the prepared aliment into its passage towards the stomach, and for carrying it along that passage. In the mean time, and within the same cavity, is going on other business wholly different, that of respiration and of speech. In addition, therefore, to all that has been mentioned, we have a passage opened from this same cavity of the mouth into the lungs for the admission of air, for the admission of air exclusively of every other substance; we have muscles, some in the larynx, and, without number, in the tongue, for the purpose of modulating that air in its passage, with a variety, a compass, and a precision of which no other musical instrument is capable; and, lastly, we have a specific contrivance for dividing the pneumatic part from the mechanical, and for preventing one set of functions from interfering with the other. The mouth, with all these intentions to serve, is a single cavity; is one machine, with its parts neither crowded nor confined, and each unembarrassed by the rest.'" It should be added, the mouth is also the immediate seat of one of the senses, and is in intimate communication with a second sense; both these senses are always excited while the principal business performed by the machine is carried on, and are necessarily excited by the very working of the machine, and the sensations induced in the natural and sound state of the apparatus are invariably pleasurable.

611. The food is delivered by the pharynx to the esophagus (fig. CLIII. 12), a tube composed partly of membrane and partly of muscle. Its muscular fibres consist of a double layer, an external and an internal layer; the external has a longitudinal direction; the internal describes portions of a circle around the tube. By the contraction of the longitudinal fibres the length, and by the contraction of the circular fibres, the diameter of the tube is diminished. Cellular membrane envelopes these layers of fibres externally, and mucous membrane covers them internally. When the tube is contracted, the mucous membrane is disposed in folds, which disappear when it is dilated, and these folds allow of the expansion of the tube without injury to the delicate tissue that lines it. The food passes slowly along the esophagus urged towards the stomach, not by its own gravity, but by a force exerted upon it by the tube itself, chiefly by the contraction of its circular fibres. Delivered at length to the stomach,

the food is incapable of returning into the esophagus in consequence of the oblique direction in which the esophagus enters the stomach, the obliquity of its entrance serving the office of a valve.

612. The stomach is a bag of an irregular oval shape (fig. CLXVI.), capable, in the adult, of containing about three pints. It is placed transversely across the upper part of the abdomen (fig. LX. 7). It occupies the whole epigastric (fig. CV. 3), and the greater part of the left hypochondriac regions (fig. CVII. 3). Above, it is in contact with the diaphragm, the arch of

Fig. CLXVI.— *View of the Stomach with its Muscular Coats displayed.*

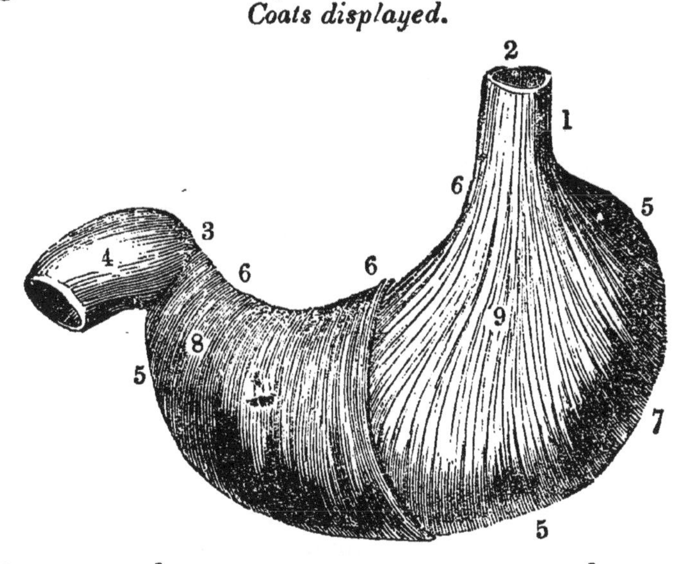

1. The esophagus terminating in the stomach. 2. The cardiac orifice. 3. The pylorus. 4. The commencement of the duodenum. 5. The large curvature of the stomach. 6. The small curvature. 7. The large extremity. 8. The small extremity. 9. The longitudinal muscular fibres. 10. The circular muscular fibres.

which extends over it (fig. LX. 7, b); below with the intestines (fig. LX. 8, 9), on the right side with the liver (fig. LX. 6), and on the left side with the spleen (fig. CLXVIII. 5).

613. Into the left extremity, which is much larger and considerably higher than the right

Fig. CLXVII. *Internal View of the Stomach and Duodenum.*

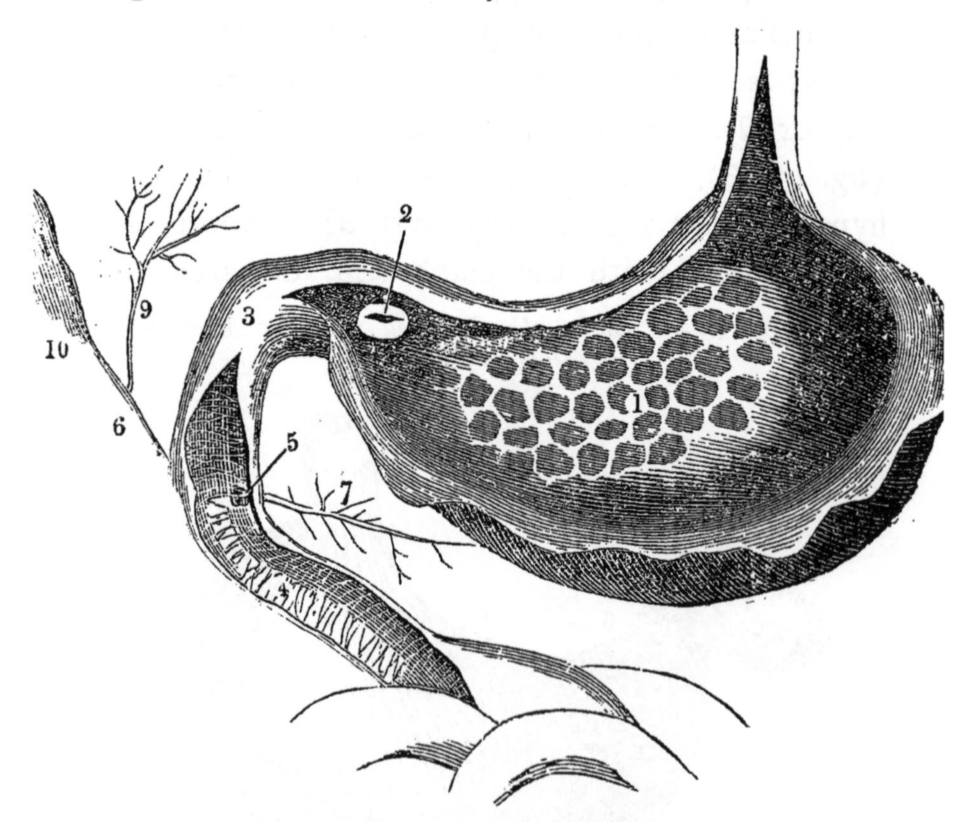

1. Mucous membrane, forming the rugæ. 2. Pyloric orifice opening into the duodenum. 3. Duodenum. 4. Interior of the duodenum, showing the valvulæ conniventes. 5. Termination of, 6. The biliary or choledoch duct. 7. Pancreatic duct, terminating at the same point as the choledoch duct. 8. Gall-bladder removed from the liver. 9. Hepatic duct proceeding from the liver. 10. Cystic duct proceeding from the gall-bladder, forming, by its union with the hepatic, a common trunk, the choledoch.

(fig. CLXVI. 7), the esophagus opens by an aperture called the cardiac orifice (fig. CLXVI. 2). At the right extremity, a second aperture called the pyloric orifice (fig. CLXVII. 2), leads into the first intestine.

614. Between the cardiac and the pyloric orifices are two curvatures, one above, called the smaller (fig. CLXVI. 6), the other below, termed the larger curvature (fig. CLXVI. 5).

615. Like the esophagus, the stomach is composed of two layers of muscular fibres, the external longitudinal (fig. CLXVI. 9), the internal circular (fig. CLXVI. 10). By the contraction of the first the extent of the stomach, from extremity to extremity, is diminished, or the organ is shortened; by the contraction of the second the extent of the stomach, from curvature to curvature, is diminished, or the organ is narrowed. During digestion, by the contraction of these muscular fibres, the capacity of the stomach is changed alternately in both directions, whence a gentle motion is communicated to the aliment, which is thus brought in succession under the influence of the agent that acts upon it.

616. A thin but strong membrane, derived from the peritoneum, the membrane that lines the general cavity of the abdomen, forms the external tunic of the stomach; hence its outer covering is called the peritoneal coat.

617. The inner or mucous coat (fig. CLXVII. 1),

a direct continuation of the lining membrane of the esophagus, is sometimes called also villous, on account of the minute bodies termed villi, with which every point of its internal surface is studded. It is these villi which give to this surface its pilous or velvety appearance.

618. The mucous coat is far more extensive than the other two, in consequence of its being plaited into a number of folds (fig. CLXVII. 1), termed rugæ, which are so disposed as to present the appearance of a net-work. The object of the rugæ is to enlarge the space for the expansion of

Fig. CLXVIII.—*View of the Vascular connexion between the Stomach, Liver, Spleen, and Pancreas.*

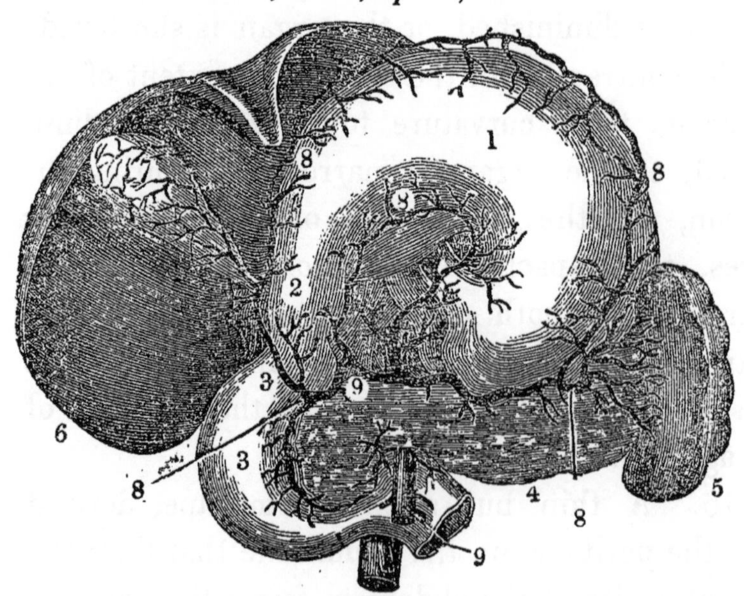

1. Stomach raised to exhibit its posterior surface, 2. Pylorus. 3. Duodenum. 4. Pancreas. 5. Spleen. 6. Undersurface of the liver. 7. Gall-bladder, in connexion with the liver. 8. Large vessels proceeding from. 9. A common trunk to supply the liver, gall-bladder, stomach, duodenum, pancreas, and spleen.

blood-vessels and nerves, and to admit of the occasional distension of the organ without injury to the delicate tissues of which it is composed.

619. Immediately beneath the mucous coat are the mucous follicles which secrete the mucous fluid by which the inner surface of the organ is defended. These glandular bodies are extremely numerous, and vary considerably in diameter. The largest are towards the great extremity, the smaller towards the pylorus.

620. Altogether different from the mucous secretion is another fluid, which also flows from the mucous surface, termed the gastric or the digestive juice, from its being the principal agent in the digestive process. By some anatomists the gastric juice is supposed to be secreted by minute glands placed between the mucous and the muscular coats, provided with ducts which pierce the mucous coat, and which bear their fluid into the stomach precisely as the salivary glands carry the saliva into the mouth. It is certain that this is the case with some animals, as in certain birds, the ostrich for example, in which glands of considerable magnitude, with ducts large enough to be visible, are seen to pour the digestive fluid into the stomach. But as no such glands have been discovered in the human stomach, it is generally conceived that in man the gastric juice is secreted by minute arteries expanded upon the villi.

621. All around the pyloric orifice (fig. CLXVII. 2)

Fig. CLXIX.

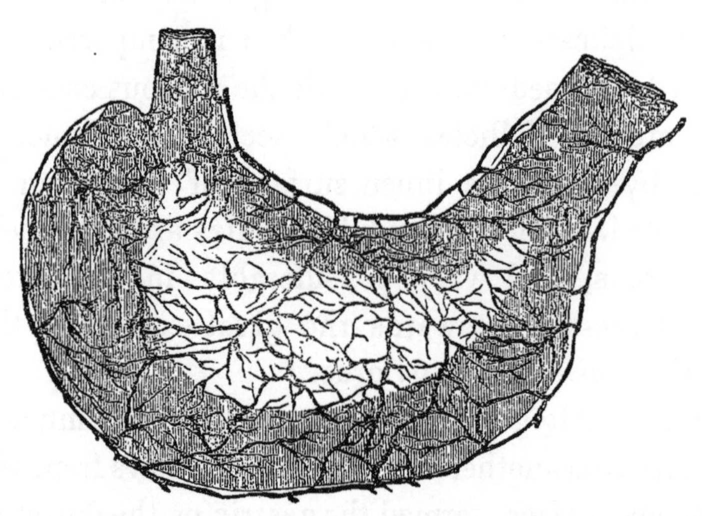

View of the stomach, showing the number and magnitude of its blood-vessels, and the mode of their distribution.

is placed a thick, strong, and circular muscle (fig. CLXVII. 2), termed, from its office, pylorus. It is about four times the thickness of the muscular coat of the stomach, and presents the appearance of a prominent and even projecting band (fig. CLXVII. 2). From the frequent action of its fibres, the pylorus often looks as if a piece of pack-thread had been tied around it (fig. CLXVI. 3). Its office is, by the contraction of its fibres, to guard and close the opening from the stomach until the aliment has been duly acted upon by the digestive fluid.

622. The quantity of blood sent to the stomach is greater than is spent upon any other organ except the brain. The vessels of the stomach (fig.

CLXIX.) form two distinct layers, of which the external is distributed to the peritoneal and muscular coats, while the internal, after ramifying on the fine cellular tissue which unites the muscular and mucous tunics, penetrates the mucous coat, and is spent upon the villi, where it forms an exquisitely-delicate net-work. There is, moreover, an intimate vascular connexion between the spleen, pancreas and liver, and the stomach (fig. CLXVIII. 8, 9). The arteries which supply all these organs spring from a common trunk, and there is the freest communication between them by anastomosing branches.

623. Equally abundant is its supply of nerves, some of which are derived from the organic or non-sentient system, and others from the animal or sentient system. The organic nerves are spread out in countless numbers upon the great trunks of the arteries, so as to give them a complete envelope (fig. CLXX. 3); these nerves, never quitting the arteries, accompany them in all their ramifications, and the fibril of the nerve is ultimately lost upon the capillary termination of the artery. It is by these organic nerves that the stomach is enabled to perform its organic functions, which, for the reason assigned (vol. i. p. 82), is placed beyond volition, and is without consciousness. By the nerves derived from the sentient system which mingle with the organic (fig. XVI.), the function of nutrition is brought into relation with the percipient mind, and is

Fig..CLXX.—*View of the Organic Nerves of the Stomach.*

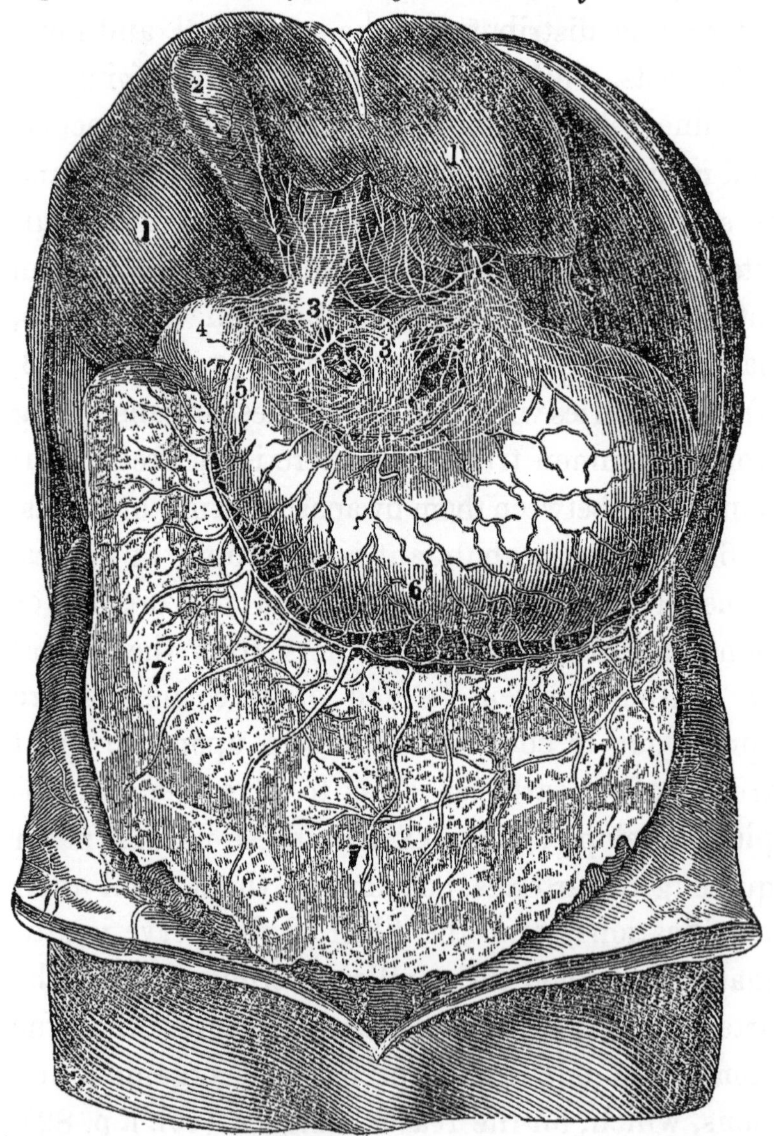

1. Under surface of the liver turned up, to bring into view the anterior surface of the stomach. 2. Gall bladder. 3. Organic nerves enveloping the trunks of the blood-vessels. 4. Pyloric extremity of the stomach and commencement of the duodenum. 5. Contracted portion of the pylorus. 6 Situation of the hour-glass contraction of the stomach, here imperfectly represented. 7. Omentum.

made part of our sentient nature. By the commixture of these two sets of nerves, derived from these two portions of the nervous system, though we have no *direct* consciousness of the digestive process—consciousness ceasing precisely at the point where the agency of volition stops (vol. i. p. 82, et seq.), yet pleasurable sensation results from the due performance of the function. Hence the feeling of buoyancy, exhilaration, and vigour, the pleasurable consciousness to which we give the name of health, when the action of the stomach is sound: hence the depression, listlessness, and debility, the painful consciousness which we call disease, when the action of the stomach is unsound: hence, too, the influence of the mental state over the organic process; the rapidity and perfection with which the stomach works when the mind is happy—when the repast is but the occasion and accompaniment of the feast of reason and the flow of soul; the slowness and imperfection with which the stomach works when the mind is harassed with care struggling against adverse events; or is in sorrow and without hope; when the friend that sat by our side, and with whom we were wont to take sweet counsel, is gone, and therefore gone that which made it life to live.

624. Renovation is the primary and essential office of the stomach, and its organic nerves enable it to supply the ever-recurring wants of the

system. Gratification of appetite is a secondary and subordinate office of the stomach, and its sentient nerves enable it to produce the state of pleasurable consciousness when its organic function is duly performed. By the double office thus assigned it, the stomach is rendered what Mr. Hunter named it, the centre of sympathies.

625. From the whole length of the great arch of the stomach, and partly also from the commencement of the duodenum (fig. CLXX.), the peritoneal coat of the stomach is produced, forming a thin, delicate membranous bag, called the omentum, or cawl (fig. CLXX. 7). The omentum extends from the great arch of the stomach to below the umbilicus, and completely covers a large portion of the anterior surface of the abdominal viscera (fig. CLXX. 7). Between the two fine membranous layers of which it is composed is contained a quantity of fat, of which substance it serves as a reservoir, and by the transudation of which it appears to lubricate the intestines, and to assist in preventing their accretion.

626. The food, on reaching the stomach, does not occupy indifferently any portion of it, but is arranged in a peculiar manner always in one and the same part. If the stomach be observed in a living animal, or be inspected soon after death, it is seen that about a third of its length towards the pylorus is divided from the rest by the contraction of the circular fibries called the hour-glass con-

traction (fig. CLXX. 6). The stomach is thus divided into a cardiac and a pyloric portion (fig. CLXX. 6). The food, when first received by the stomach, is always deposited in the cardiac portion, and is there arranged in a definite manner. The food first taken is placed outermost, that is, nearest the surface of the stomach; the portion next taken is placed interior to the first, and so on in succession, until the food last taken occupies the centre of the mass. When new food is received before the old is completely digested, the two kinds are kept distinct, the new being always found in the centre of the old.

627. Soon after the food has been thus arranged, a remarkable change takes place in the mucous membrane of the stomach. The blood-vessels become loaded with blood; its villi enlarge, and its cryptæ, the minute cells between the rugæ, overflow with fluid. This fluid is the gastric juice, which is secreted by the arterial capillaries now turgid with blood. The abundance of the secretion, which progressively increases as the digestion advances, is in proportion to the indigestibility of the food, and the quietude of the body after the repast.

628. In the food itself no change is manifest for some time; but at length that portion of it which is in immediate contact with the surface of the stomach begins to be slightly softened. This softening slowly but progressively increases until

the texture of the food, whatever it may have been, is gradually lost; and ultimately the most solid portions of it are completely dissolved.

629. When a portion of food thus acted on is examined, it presents the appearance of having been corroded by a chemical agent. The white of a hard-boiled egg looks exactly as if it had been plunged in vinegar or in a solution of potass. The softened layer, as soon as the softening is sufficiently advanced, is, by the action of the muscular coat of the stomach, detached, carried towards the pylorus, and ultimately transmitted to the duodenum; then another portion of the harder and undigested food is brought into immediate contact with the stomach, becomes softened in its turn, and is in like manner detached; and this process goes on until the whole is dissolved.

630. The solvent power exerted by the gastric juice is most apparent when the stomach of an animal is examined three or four hours after food has been freely taken. At this period the portion of the food first in contact with the stomach is wholly dissolved and detached; the portion subsequently brought into contact with the stomach is in the process of solution, while the central part remains very little changed.

631. The dissolved and detached portion of the food, from every part of the stomach flows slowly but steadily beyond the hour-glass contraction, or towards the pyloric extremity (626), in

which not a particle of recent or undissolved food is ever allowed to remain. The fluid, which thus accumulates in this portion of the stomach, is a new product, in which the sensible properties of the food, whatever may have been the variety of substances taken at the meal, are lost. This new product, which is termed chyme, is an homogeneous fluid, pultaceous, greyish, insipid, of a faint sweetish taste, and slightly acid.

632. As soon as the chyme, by its gradual accumulation in the pyloric extremity amounts to about two or three ounces, the following phenomena take place.

633. First, the intestine called duodenum, the organ immediately continuous with the stomach, contracts. The contraction of the duodenum is propagated to the pyloric end of the stomach. By the contraction of this portion of the stomach, the chyme is carried backwards from the pyloric into the cardiac extremity, where it does not remain, but quickly flows back again into the pyloric extremity, which is now expanded to receive it. Soon the pyloric extremity begins again to contract; but now the contraction, the reverse of the former, is in the direction of the duodenum; in consequence of which, the chyme is propelled towards the pylorus. The pylorus, obedient to the demand of the chyme, relaxes, opens, and affords to the fluid a free passage into the duodenum. As soon as the whole of the duly prepared

chyme has passed out of the stomach, the pylorus closes, and remains closed, until two or three ounces more are accumulated, when the same succession of motions are renewed with the same result; and again cease to be again renewed, as long as the process of chymification goes on.

634. When the stomach contains a large quantity of food, these motions are limited to the parts of the organ nearest the pylorus; as it becomes empty, they extend further along the stomach, until the great extremity itself is involved in them. These motions are always strongest towards the end of chymification.

635. The stomach during chymification is a closed chamber; its cardiac orifice is shut by the valved entrance of the esophagus, and its pyloric orifice by the contraction of the pylorus.

636. The rapidity with which the process of chymification is carried on is different according to the digestibility of the food, the bulk of the morsels swallowed, the quantity received by the stomach, the constitution of the individual, the state of the health, and above all, the class of the animal, for it is widely different in different classes. In the human stomach in about five hours after an ordinary meal the whole of the food is probably converted into chyme.

637. The great agent in performing the process of chymification is the gastric juice. The evidence of this is complete for,

1. As soon as the food enters the stomach a large quantity of blood is determined to the arteries, which secrete the gastric juice (627); and this fluid continues to be poured into the stomach in great abundance during the whole time the process goes on.

2. The solvent power of this fluid is demonstrated by the fact that it sometimes dissolves the stomach itself, when death takes place suddenly during the act of digestion in a sound and vigorous state of the digestive organs.

3. On introducing into the stomach alimentary substances inclosed in metallic balls perforated with holes, or in pieces of porous cloth, it is found, on removing these bodies from the stomach, after a certain time, that the alimentary substances contained in them are as completely digested as if they had been in actual contact with the surface of the stomach; the metallic ball and the cloth remaining wholly unchanged. This experiment, which has been often performed with the same uniform result, was the first that led to the discovery of the true nature of the digestive process.

4. Though it be impossible to imitate out of the stomach all the circumstances under which the food is placed within it, yet, on procuring gastric juice from the stomachs of various animals, and mixing it with different alimentary substances, it is found not only to dissolve them, but to convert them into a pultaceous mass, closely resembling chyme.

Gastric juice thus procured was put into a glass tube with boiled beef, which had been masticated; the tube was then hermetically sealed, and exposed near the fire to a uniform heat: by the side of this tube was placed another, containing the same quantity of flesh immersed in water. In twelve hours, the flesh in the tube containing the gastric juice began to lose its fibrous structure; in thirty-five hours it had nearly lost its consistence, being reduced to a soft homogeneous pultaceous mass. It experienced no further change during the two following days. On the other hand, the flesh that had been immersed in water was putrid in sixteen hours.

638. Since alimentary substances under the action of the stomach present precisely the appearance exhibited by bodies exposed to the influence of chemical agents, it appears that the gastric juice not only dissolves the food, but dissolves it by a chemical agency. Its action bears no proportion to the mechanical texture of bodies, nor to any of their physical properties. It acts upon the densest membrane, dissolves even bone itself; and yet produces no effect on other substances of the most tender and delicate texture. On the skin of fruit, on the finest fibre of flax and cotton, it is incapable of making the slightest impression. In this selection of substances it perfectly resembles a chemical agent acting by chemical affinity. On certain substances its

action is unquestionably of a chemical nature. It occasions the coagulation of albuminous fluids; it prevents the accession of putrefication; it stops the process after it has commenced. From the whole, it follows that the food in the stomach is converted into chyme by the agency of a fluid secreted by the inner surface of the stomach, and that this change is effected by a proper chemical action.

639. It had been long ascertained that the gastric juice contains an uncombined acid, and that if carbonate of lime be placed in a tube and introduced into the stomach, the carbonate is dissolved just as if it were put into weak vinegar. Several years ago, it was discovered by Dr. Prout that this free acid is muriatic acid. Some time after the publication of Dr. Prout's experiments, Chevreul and Leuret and Lassaigne in France obtained different results; but Tiedemann and Gmelin, professors in the university of Heidelberg, in an extended series of experiments, arrived at precisely the same conclusion as the English physiologist, and apparently without any previous knowledge of the researches of the latter. Tiedemann and Gmelin state, as the result of their experiments, that the clear ropy fluid, or the gastric juice obtained from the stomach some time after it had been without food, is nearly or entirely destitute of acidity; that the presence of food, or indeed of any stimulus to the mucous membrane, causes the gastric juice to become dis-

tinctly acid; that this acidity increases according to the indigestibility of the food; that the quantity of acid poured out is very copious; that it consists partly of muriatic and partly of acetic acid; and that both these acids are efficient agents in the process of digestion. Dr. Prout, who had also recognised the presence of acetic acid, is of opinion that its formation is an accidental occurrence not necessary to digestion nor conducive to it; but is either derived from the aliment, or is the result of irritation or disease. He contends that the muriatic acid is the efficient digestive agent.

640. The still more recent experiments of Braconnot appear to have set this matter at rest, and to have proved, beyond all controversy, that the stomach, when stimulated by the presence of food or other foreign agents, possesses the power of secreting free muriatic acid in great quantity; and that it is by this acid that the solution of the food is effected. It is even found that muriatic acid is capable of digesting alimentary substances out of the body. It had been long known, that if meat and gastric juice be inclosed in a tube and kept at the temperature of the human body, a product is obtained closely resembling chyme (637.4). M. Blondelot, a physician at Nancy, has recently shown that the same result may be obtained by the digestion of the muscular fibre, in dilute muriatic acid. In both cases the muscular fibre retains its form and its original fibrous texture; but on the

slightest motion it divides into an insoluble mass, perfectly homogeneous and similar to the chyme of the stomach;* a very close approximation to the actual digestive process, more especially when it is considered that it is not possible to imitate out of the stomach several circumstances materially influencing chemical action under which the food is placed within the stomach.

641. Muriatic acid, the chemical agent by which the stomach dissolves the food, is probably obtained from the muriate of soda (common salt) contained in the blood. The soda, the basis of the salt, would appear to be retained in the blood, to preserve the alkaline condition essential to the maintenance of the sound constitution of the blood, while the muriatic acid, disengaged from the soda in the process of secretion, is poured into the stomach to act upon the food.

642. A remarkable confirmation of the correctness of the general conclusions to which observation and experiment had thus enabled physiologists to arrive, is afforded by the case of a young soldier in the American army, of the name of Alexis St. Martin, who received a wound on the left side by the accidental discharge of a musket. The charge, which consisted of duck shot, and which was received at the distance of one yard from the muzzle of the gun, entered the side posteriorly in

* Dr. R. Thomson, British Annals of Medicine, No. 13.

an oblique direction, forward and inward; blew off the integument and muscles to the size of a man's hand; fractured and carried away the anterior half of the sixth rib; fractured the fifth rib; lacerated the lower portion of the left lobe of the lungs; lacerated the diaphragm, and perforated the stomach.

643. Violent fever and extensive sloughing of the parts injured ensued, and the life of the young man was often in jeopardy, but he ultimately recovered. At the distance of about a year from the date of the accident, the injured parts had all become sound, with the exception of the perforation into the stomach, which never closed, but left an aperture permanently open, two inches and a half in circumference. This aperture was situated about three inches to the left of the cardia, near the left superior termination of the great curvature. For some time the food could be retained only by constantly wearing a compress and bandage; but at length a small fold of the mucous coat of the stomach appeared, which increased until it completely filled the aperture and acted as a valve, so as effectually to prevent any efflux from within, while it admitted of being easily pushed back by the finger from without: when the stomach was nearly empty, it was easy to examine its cavity to the depth of five or six inches by artificial distension; but, when entirely empty, the stomach was always contracted on

itself, and the valve generally forced through the orifice, together with a portion of the mucous membrane equal in bulk to a hen's egg.

644. It chanced that the admirable opportunity thus afforded of bringing the process of digestion, as far as it is carried on in the stomach, under the cognizance of sense, occurred to an observant and philosophical mind, and it was not lost.* The following are some of the curious and instructive phenomena observed.

645. The inner coat of the stomach, in its natural and healthy state, is of a light or pale pink colour, varying in its hues according to its full, or empty state. It is of a soft or velvet-like appearance (617), and is constantly covered with a very thin transparent, viscid mucus, lining the whole interior of the organ (619).

646. Immediately beneath the mucous coat appear small spheroidal, or oval-shaped glandular bodies, from which the mucous fluid appears to be secreted (619).

647. By applying aliment or other irritants to the internal coat of the stomach, and observing the effect through a magnifying glass, innumerable minute lucid points, and very fine nervous or vascular papillæ are seen arising from the villous membrane, and protruding through the mucous

* Experiments and Observations on the Gastric Juice, and the Physiology of Digestion. By W. Beaumont, M.D., Surgeon in the U. S. Army. Boston. 1834.

coat, from which distils a pure, limpid, colourless, slightly viscid fluid (620). This fluid, thus excited, is invariably distinctly acid (639, *et seq.*). The *mucus* of the stomach is less fluid, more viscid or albuminous, semi-opaque, sometimes a little saltish, and does not possess the slightest character of acidity (619). On applying the tongue to the mucous coat of the stomach in its empty, unirritated state, no acid taste can be perceived. When food or other irritants have been applied to the villous membrane and the gastric papillæ excited, the acid taste is immediately perceptible. The invariable effect of applying aliment to the internal, but exposed part of the gastric membrane, is the exudation of the solvent fluid from the papillæ. Though the aperture of these vessels cannot be seen even with the assistance of the best microscopes, yet the points from which the fluid issues are clearly indicated by the gradual appearance of innumerable very fine lucid specks rising through the transparent mucous coat, and seeming to burst and discharge themselves upon the very points of the papillæ, diffusing a limpid thin fluid over the whole interior gastric surface.

648. The fluid so discharged is absorbed by the aliment in contact; or collects in small drops, and trickles down the sides of the stomach to the more depending parts, and there mingles with the food, or whatever else may be contained in the gastric cavity. This fluid, the efficient cause of diges-

tion, the true gastric juice is secreted only when it is needed; it is not accumulated in the intervals of digestion, to be ready for the next meal; it is seldom if ever discharged from its proper secreting vessels, except when excited by the natural stimulus of aliment, the mechanical irritation of tubes, or other excitants. When aliment is received, the juice is given out in exact proportion to its requirements for solution, except when more food has been taken than is necessary for the wants of the system.

649. On collecting this fluid, which it was easy to obtain, it was found to be transparent, inodorous, saltish, and acidulous to the taste; it consisted of water, containing free muriatic and acetic acids, phosphates and muriates, with bases of potass, soda, magnesia, and lime, together with an animal matter soluble in cold, but insoluble in hot water.

650. When a portion of liquid aliment, as a few spoonsful of soup, were introduced into the stomach at the external orifice, the rugæ (fig. CLXVII. 1) immediately closed gently upon it; gradually diffused it through the gastric cavity, and prevented the entrance of a second quantity till this diffusion was effected; then relaxation again took place, and admitted of a further supply. When solid food was introduced in the same manner, either in large pieces or finely divided, the same gentle contraction and grasping motions were

excited, and continued from fifty to eighty seconds, so as to prevent more from being introduced, without considerable force till the contraction was at an end.

651. When the position of the body was such that the cardiac portion of the stomach was brought into view, and a morsel of food was swallowed in the natural mode, a similar contraction of the stomach, and closing of its fibres upon the bolus was invariably observed to take place; and till this was over, a second morsel could not be received without a considerable effort. Hence, in addition to the other purposes accomplished by mastication, insalivation, and deglutition, it is probable that these operations answer the further use of duly regulating the time for the admission of successive portions of the food into the stomach.*

652. On watching the phenomena that take place on the contact of a portion of food with the stomach, the circumstances described (627) are seen; the change in the mucous coat from a pale pink to a deep red colour, in consequence of the enlargement of the blood-vessels and their admission of a greatly increased number of red particles; the undulating motion of the stomach, in conse-

* See Dr. Andrew Combe on the Physiology of Digestion, in whose work a full detail of this instructive case is given. See also Mayo's Outlines of Physiology 4th Edit. Appendix.

quence of the contraction of its muscular fibres, excited by the stimulus of food; the distillation of the gastric juice from the enlarged and excited papillæ; the continuous flow of this fluid until the complete solution of the food, when food is present; and, on the contrary, the cessation of this discharge in a short time when it is produced by a mechanical irritant, as the bulb of a thermometer, although at first the gastric juice distil from the papillæ, from the contact of such an irritant, just as when excited by the contact of food.

653. On collecting the gastric juice and placing it in contact with an alimentary substance out of the stomach, its solution takes place more slowly, but not less completely, than when retained in the stomach. An ounce of this fluid was placed in a vial with a piece of boiled, recently salted beef, weighing three drachms; the vial was then tightly corked, and immersed in water, raised to the temperature of 100°, previously ascertained to be the ordinary heat of the stomach. In forty minutes the process of solution had commenced on the surface of the beef. In fifty minutes the texture of the beef began to loosen and separate. In sixty minutes an opaque and cloudy fluid was formed. In one hour and a half the muscular fibres hung loose and unconnected, and floated about in shreds in the more fluid matter. In three hours the muscular fibres had diminished about one half. In five hours only a few remained undissolved. In

seven hours the muscular texture was no longer apparent; and in nine hours the solution was completed.

654. At the commencement of this experiment a piece of the same beef of equal weight and size was suspended within the stomach by means of a string. On examining this portion of beef at the end of half an hour, it was found to present precisely the same appearance as the piece in the vial; but on the removal of the string at the end of an hour and a half the beef had been completely dissolved, and had disappeared, making a difference of result in point of time of nearly seven hours. In both, the solution began on the surface, and agitation accelerated its progress by removing the external coating of chyme as fast as it was formed.

655. An ordinary dinner having been taken, consisting of boiled salted beef, bread, potatoes, and turnips, with a gill of pure water for drink, a portion of the contents of the stomach was drawn off into an open-mouthed vial, twenty minutes after the meal. The vial was placed in a water-bath, maintained steadily at a temperature of 100°. It was continued in this temperature for five hours. At the end of that time the whole contents of the vial were dissolved. On comparing the solution with an equal quantity of chyme taken from the stomach, little difference could be distinguished between the two fluids, excepting that it was manifest that the digestive process had proceeded some-

what more rapidly in, than out of the stomach. The food, in this experiment, after having remained in contact with the stomach for the space of twenty minutes, had imbibed a sufficient quantity of gastric juice to complete its solution.

656. Fifteen minutes after half a pint of milk had been introduced into the stomach, it presented the appearance of a fine loosely-coagulated substance mixed with a semi-transparent whey-coloured fluid. A drachm of warm gastric juice poured into two drachms of milk at a temperature of 100°, produced a precisely similar appearance in twenty minutes. In another experiment, when four ounces of bread were given with a pint of milk, the milk was coagulated and the bread reduced to a soft pulp in thirty minutes, and the whole was completely digested in two hours.

657. When the albumen or white of two eggs was swallowed on an empty stomach, small white flakes began to be seen in about ten or fifteen minutes, and the mixture soon assumed an opaque whitish appearance. In an hour and a half the whole had disappeared. Two drachms of albumen mixed with two of gastric juice out of the stomach underwent precisely the same changes, but in a somewhat longer time.

658. Dr. Beaumont's observations are adverse to the opinion, founded on numerous experiments, that the food is arranged in the stomach in a definite manner, and that a distinct line of separation

exists between old and new food (626). In the human stomach, according to the subject of these experiments, the ordinary course and direction of the food are first from right to left along the small arch, and thence through the large curvature from left to right. The bolus as it enters the cardia turns to the left, passes the aperture, descends into the splenic extremity, and follows the great curvature towards the pyloric end. It then returns in the course of the smaller curvature, makes its appearance again at the aperture, in its descent into the great curvature, to perform similar revolutions. These revolutions are completed in from one to three minutes. They are probably induced in a great measure by the circular or transverse muscles of the stomach (615), as is indicated by the spiral motion of the stem of the thermometer, both in descending to the pyloric portion, and in ascending to the splenic. These motions are slower at first than after chymification has considerably advanced. The whole contents of the stomach, until chymification be nearly complete, exhibit a heterogeneous mass of solids and fluids, hard and soft, coarse and fine, crude and chymified; all intimately mixed, and circulating promiscuously through the gastric cavity like the mixed contents of a closed vessel, gently agitated or turned in the hand.

659. In attempting to pass a long glass thermometer through the aperture into the pyloric portion

of the stomach, during the latter stages of digestion, a forcible contraction is perceived at the point of the hour-glass contraction of the stomach, and the bulb is stopped. In a short time there is a gentle relaxation, when the bulb passes without difficulty, and appears to be drawn quite forcibly, for three or four inches, towards the pyloric end. It is then released, and forced back, or suffered to rise again, at the same time giving to the tube a circular or rather a spiral motion, and frequently revolving it quite over. These motions are distinctly indicated and strongly felt in holding the end of the tube between the thumb and finger; and it requires a pretty forcible grasp to prevent it from slipping from the hand, and being drawn suddenly down to the pyloric extremity. When the tube is left to its own direction at these periods of contraction, it is drawn in, nearly its whole length, to the depth of ten inches; and when drawn back requires considerable force, and gives to the fingers the sensation of a strong suction power, like drawing the piston from an exhausted tube. This ceases as soon as the relaxation occurs, and the tube rises again, of its own accord, three or four inches, when the bulb seems to be obstructed from rising further; but if pulled up an inch or two through the stricture, it moves freely in all directions in the cardiac portions, and mostly inclines to the splenic extremity, though not disposed to make its exit at the aperture. These peculiar motions and contractions continue until the stomach is perfectly empty, and not a particle of food or chyme remains, when all becomes quiescent again.

660. The chambers in which the remaining part of the digestive process is carried on are much less accessible, and no such favourable opportunity as that enjoyed by Dr. Beaumont has occurred of rendering their operations manifest to the eye. Nevertheless, the researches of physiologists have succeeded in disclosing, with almost equal exactness and certainty, the successive changes which the food undergoes even in these more hidden organs, that admit of no exposure during life without extreme danger.

661. The chyme on passing through the pylorus is received into a chamber (fig. CLXVII. 3) which forms the first portion of the small intestines. The small intestines, taken together, constitute a tube about four times the length of the body. This tube is conical, the base of the cone being towards the pylorus, and its apex at the valve of the colon, where the small intestines terminate in the large. From the pylorus to the valve of the colon the small intestines diminish in capacity, in thickness, in vascularity, in the size of the villi, and in the depth and number of the valvulæ conniventes.

662. The first portion of the small intestine is termed the duodenum (fig. CLXVII. 3). It is about twelve inches in length, and, unlike the stomach, which is capable of considerable motion, it is

closely tied down to the back by the peritoneum, which imperfectly covers it. The rest of the small intestine is divided into two portions—the

Fig. CLXXI.

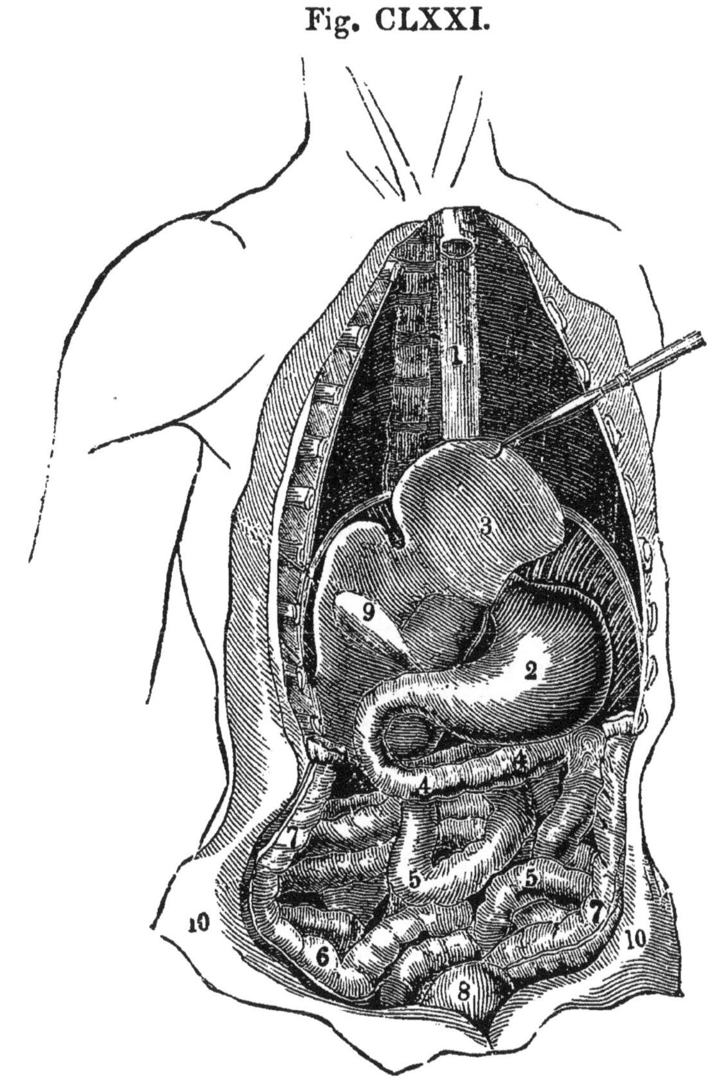

1. Esophagus. 2. Stomach. 3. Liver raised, showing the under surface. 4. Duodenum. 5. Small intestines, consisting of—6. Jejunum and ilium. 7. Colon. 8. Urinary bladder. 9. Gall bladder. 10. Abdominal muscles di vided and reflected.

upper two-fifths of which are termed jejunum, and the three lower ilium.

663. The duodenum, the chamber which receives the chyme from the pylorus, is a second stomach, which carries on the process commenced in the first. It is assisted in the performance of its function by two organs of considerable magnitude, the pancreas and the liver.

664. The pancreas is a conglomerate gland (fig. CLXXII. 5), of an elongated form, placed in the epigastric region, lying transversely across it, immediately behind the stomach (fig. CLXXII. 1), and resting upon the spinal column (fig. CLXXII. 5). Its right extremity is attached to the duodenum (fig. CLXXII. 9), and its left to the spleen (fig. CLXXII. 4). In external appearance it resembles the salivary glands, but it is of much larger size, and its weight, from four to six ounces, is three times greater than that of all the salivary glands together. It secretes a peculiar fluid called the pancreatic juice, which is carried into the duodenum by a tube named the pancreatic duct (fig. CLXVII. 7), which opens into the duodenum about four or five inches from its pyloric end (fig. CLXVII. 2).

665. The liver, the largest and heaviest gland in the body, weighing about four pounds, is placed chiefly in the right hypochondriac region (fig. CLXXI. 3); but a portion of it extends transversely across the epigastric, into the left hypochondriac

Fig. CLXXII.

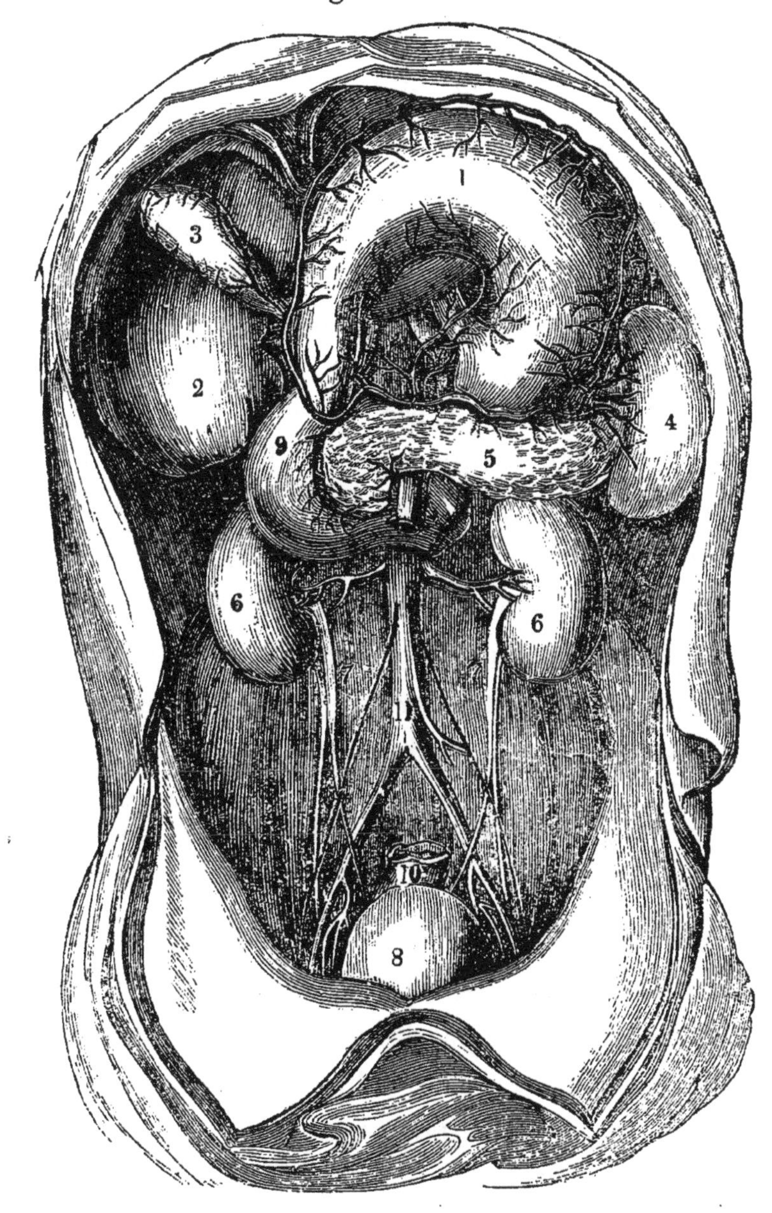

1. Stomach raised. 2. Under surface of liver. 3. Gall bladder. 4. Spleen. 5. Pancreas. 6. Kidneys. 7. Ureters. 8. Urinary bladder. 9. Portion of intestine called duodenum. 10. Portion of intestine called rectum. 1. Aorta.

region (figs. CV. and CVII. 3). Its upper surface is in contact with the diaphragm (fig. LX. 6, b); its under surface with the pyloric extremity of the stomach (fig. LX. 7), and its margin can be felt under the edges of the ribs of the right side.

666. It has been stated (473, 1.) that the fluid secreted by the liver, unlike that formed by any other organ of the body, is elaborated from venous blood, derived from the veins of the digestive organs, and that these veins uniting together, form a common trunk called the vena portæ, which penetrates the liver and ramifies through it in the manner of an artery. Galen long ago compared this venous system to a tree whose roots are dispersed in the abdomen, and its branches spread out through the liver. Two comparatively small arteries, called the hepatic, nourish the liver; the ultimate divisions of these arteries likewise terminate in the vena portæ. The ultimate branches of the vena portæ terminate partly in a system of veins, called the hepatic, which like ordinary veins return the blood to the right side of the heart; and partly in a system of tubes, termed the biliary ducts, which contain the fluid secreted by the capillary branches of the vena portæ. This fluid is the bile. The biliary ducts uniting from all parts of the liver by innumerable branches, at length form a single trunk termed the hepatic duct (fig. CLXVII. 9), which carries the bile partly to the gall bladder (fig. CLXVII. 8)

by a duct called the cystic (fig. CLXVII. 10), and partly to the duodenum (fig. CLXVII. 3) by a duct named the choledoch (fig. CLXVII. 6), a common trunk formed by the union of the cystic with the hepatic (fig. CLXVII. 10 and 9). The choledoch duct opens into the duodenum at the same point as the pancreatic (fig. CLXVII. 7), and generally by a common orifice.

667. The duodenum, on receiving the chyme from the stomach, transmits it slowly along its surface. The kind of motion by which the chyme is borne along the surface of the duodenum is perfectly analogous to that by which it is transmitted from the stomach to the duodenum, irregular, sometimes in one direction, and sometimes in another, at one time commencing in one part of the organ, at another time in another, always slow, but ultimately progressive.

668. As the chyme slowly advances through the upper part of the duodenum, the biliary and the pancreatic juices slowly distil into the lower portion of the organ. The bile is seen to exude from the choledoch duct, not continually, but at intervals, a drop appearing at the orifice, and diffusing itself over the neighbouring surface, about twice in a minute, while the flow of the pancreatic juice is still slower.

669. No appreciable change takes place in the chyme until it reaches the orifice of the choledoch duct: but as soon as it comes in contact with this

portion of the duodenum, the chyme suddenly loses its own sensible properties, and acquires those of the bile, especially its colour and bitterness. But these properties are not long retained; a spontaneous change soon takes place in the compound. It separates into a white fluid and into a yellow pulp. The white fluid is the nutritive part of the aliment; the yellow pulp is the excrementitious matter.

670. This white fluid, the proper product of the digestive process, as far as it has yet advanced, is called chyle. If any portion of oil or fat have been contained in the food, the chyle is of a milk-white colour; if not, it is nearly transparent. It is of the consistence of cream, and it bears a close resemblance to cream in its sensible properties. It differs from chyme in being of a whiter colour, more pellucid, and of a thicker consistence: it differs also in its chemical nature, for, whereas chyme is acid, chyle is alkaline.

671. Three fluids are mixed with the chyme in the duodenum, each of which contributes to the conversion of the chyme into chyle. First, the secretion of the duodenum itself, a solvent analogous to the gastric juice. Secondly, the secretion of the pancreas, a watery fluid holding in solution highly important principles, namely, a large quantity of albumen, a matter resembling casein, osmazome, and different salts. Thirdly, the secretion of the liver, a compound fluid, consisting of water, mucus, and several peculiar animal matters,

namely, resin, cholesterine, picromel, cholic acid, a colouring matter, probably salivary matter, osmazome, casein, and many salts.

672. There cannot be a question that the secretion of the duodenum has a solvent power over the chyme analogous to that of the gastric juice. Some physiologists indeed maintain that the juice poured out from the inner surface of the duodenum is as powerful a solvent as the gastric juice. It is certain that substances which have escaped chymification in the stomach undergo that process in the duodenum, and that there is the closest analogy between the action of the duodenum on the chyme and that of the stomach on the crude food

673. The pancreatic secretion adds to the chyme richly azotized animal substances, albumen, casein, osmazome (671), by which it is brought nearer the chemical composition of the blood, and prepared for its complete assimilation into it. The first addition of such assimilative matter, it has been shown, is communicated by the salivary glands, but far more important additions are now supplied from the pancreas. Hence the larger size of the pancreas and the more copious secretion of the pancreatic fluid, in herbivorous than in carnivorous animals; hence the change produced in the size of the pancreas by a long continued change in the habits of an animal; hence the smaller size of the pancreas in the wild cat, which lives wholly on animal food, than in the domestic cat, which

lives partly on animal and partly on vegetable food.

674. The bile, the most complex secretion in the body, accomplishes manifold purposes.

1. Like the pancreatic secretion, it communicates to the chyle richly azotized animal substances, picromel, osmazome, and cholic acid (671); by the combination of which with the chyme, it is brought still nearer the chemical composition of the blood. These principles are manifestly united with the chylous portion of the chyme, since they are not discoverable in its excrementitious matter.

2. Bile has the property of dissolving fat; consequently, when oily or fatty matters are contained in the food, it powerfully assists in converting these substances into chyle.

3. The excrementitious portion of the bile is highly stimulant. The contact of its bitter resin with the mucous membrane of the intestines excites the secretion of that membrane; hence the extreme dryness of the excrementitious matter when the choledoch duct of an animal has been tied; and hence the same dryness of this matter in jaundice, when the bile, instead of being conveyed by its appropriate duct into the duodenum, is taken up by the absorbents, poured into the blood, and distributed over the system.

4. The bitter resin of the bile stimulates to contraction the fibres of the muscular tunic of the intestines: by the contraction of these fibres the excre-

mentitious matter is conveyed in due time out of the body; hence the constipated state of the bowels invariably induced when the secretion of the bile is deficient, or when its natural course into the intestines is obstructed.

5. The excrementitious portion of the bile exerts an antiseptic influence over the excrementitious portion of the food during its passage through the intestines. In animals in which the choledoch duct has been tied, the excrementitious portion of the food is invariably found much further advanced in decay than in the natural state. This is also uniformly the case in the human body in proportion as the secretion of the bile is deficient, or its passage to the intestine is obstructed.

675. Such appear to be the real purposes accomplished by the bile in the process of digestion. Several uses have been assigned to it, in promoting this process, which it does not serve. Seeing the instantaneous change wrought in the chyme on its contact with the bile, it was reasonable to suppose that the main use of the bile was to convert chyme into chyle, a purpose apparently of sufficient importance to account for the immense size of the gland constructed for its elaboration. The soundness of this conclusion appeared to be established by direct experiment. Mr. Brodie placed a ligature around the choledoch duct of an animal: after the operation the animal ate as usual: on killing the animal some time after it had taken a

meal, and examining the body immediately after death, it was clear that chymification had gone on in the stomach just as when the choledoch duct was sound, but no chyle appeared to be contained either in the intestines or in the lacteals. In the lacteals there was found only a transparent fluid, which was supposed to consist of lymph and of the watery portion of the chyme. Mr. Brodie's experiments seemed to be confirmed by those of Mr. Mayo, who arrived at the conclusion, that when the choledoch duct is tied, and the animal is examined at various intervals after eating, no trace whatever of chyle is discoverable in the lacteal vessels. But these experimentalists inferred that no chyle existed in the intestines or lacteals, because there was present no fluid of a milk-white colour, a colour not essential to chyle, but dependent on the accident of oily or fatty matter having formed a portion of the food. These experiments have been repeated in Germany by Tiedemann and Gmelin, and in France by Leuret and Lassaigne, who have invariably found, after tying the choledoch duct, nearly the same chylous principles, with the exception of those derived from the bile, as in animals perfectly sound; and the English physiologists have since admitted that their German and French colaborateurs have arrived at conclusions more correct than their own.

676. The bile consists then of two different portions; a highly animalized portion, which combines

with the chyme and exalts its nature by approximating it to the condition of the blood; and an excrementitious portion, which, after accomplishing certain specific uses, is carried out of the system with the undigested matter of the food. The excrementitious portion of the bile, namely, the resin, the fat, the colouring principle, the mucus, the salts, constitute by far the largest portion of it. These constituents of the bile for the most part contain a very large proportion of carbon and hydrogen, and the reasons have been already fully stated (473, *et seq.*) which favour the conclusion that the elimination of these substances under the form of bile is one most important mode of maintaining the purity of the blood, and that the liver is thus a proper respiratory organ, truly auxiliary to the lungs. It is a beautiful arrangement, and like one of the adjustments of nature, that the bile, the formation of which abstracts from the blood so large a portion of carbon and hydrogen as to maintain the purity of the circulating mass and to counteract its putrescent tendency, acts on the excrementitious portion of the food, always highly putrescent, as a direct and powerful antiseptic; that the very matter which is eliminated on account of the putrid taint it communicates to the blood, on its passage out of the body, stops the putrefaction of the substances which have been ministering to the replenishment of the blood.

677. The chyle, thick, glutinous, and adhesive,

attaches itself with some degree of tenacity to the mucous surface of the duodenum. Nevertheless, by the successive contractions of the muscular fibres of the duodenum the fluid is slowly but progressively propelled forwards. The separation of the excrementitious matter becomes more complete, and consequently the chyle more pure as it advances, until, having traversed the course of the duodenum, it enters the second portion of the small intestines, the jejunum.

678. The jejunum, so called because it is commonly found empty, and the ilium, named from the number of its convolutions, on account of their great length, are provided with a distinct membrane to support them, and to retain them in their situation, termed the mesentery.

679. The mesentery is a broad membrane composed of two layers of peritoneum. Between these two layers, at one extremity of the duplicature, is placed the intestines, while the other extremity is attached to the spinal column. The mesentery being much shorter than the intestines, the intestines are gathered or puckered upon the membrane, by which beautiful mechanical contrivance they are held in firm and close contact with each other, yet their convolutions cannot be entangled, nor can they be shaken from their place by the sudden and often violent movements of the body. It sometimes happens, in consequence of disease, that the convolutions of the intestines are glued together by the effusion of lymph, and then the most trifling

causes are capable of producing the severest symptoms of obstruction in the bowels.

680. The internal surface of the small intestines is distinguished,

1. By the number of the mucous glands, which may be seen by a magnifying glass to consist partly of a prodigious number of the minutest follicles, not collected in groups, but equally scattered throughout; and partly of glands of a larger dimension, disposed in groups at particular parts of the canal.

2. By the increase in the number and size of the villi, of which there are about four thousand to the surface of a square inch. Like those of the stomach, the villi of the small intestine are composed of arteries, veins, nerves, and mucous ducts; but to the villi of the small intestine, in length about one-fourth of a line, there is added a new vessel, the absorbent of the chyle, the lacteal (figs. 175 and 176), so named from the milk-like chylous fluid which it contains.

3. By the great extension of the mucous coat obtained by the disposition of the membrane into the folds called valvulæ conniventes (fig. CLXXIII.). These folds, which rarely extend through the whole circle of the intestine, are often joined by communicating folds (fig. CLXXIII.). The folds are broadest in the middle, and narrowest at the extremities (fig. CLXXIII.). In general, they are about a line and a half broad. One edge of the fold is loose,

but the other is fixed to the intestine (fig. CLXXIII.). The office of these folds is, first, without in-

Fig. CLXXIII.

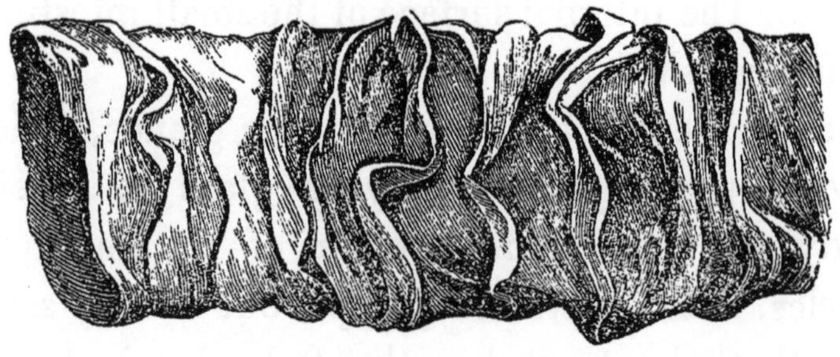

Internal view of a portion of the jejunum, showing the arrangement of the mucous membrane into valvulæ conniventes.

creasing space, to extend surface for the distribution of the villi; and, secondly, to retard the flow of the chyle, by opposing to its descent valves so constructed and disposed as, without arresting its progress, to moderate and regulate its course, in order that time may be allowed for its absorption.

681. The onward flow of the chyle through the course of the small intestines is effected by the action of the double layer of muscular fibres, the circular and the longitudinal fasciculi which compose its muscular coat (fig. CLXXIV.). The disposition of the muscular fibres of the alimentary canal in general, and of this part of it in particular, deserves special notice. The ordinary arrangement and action of muscular fibres would not have produced in this case the kind and degree of motion required. The muscular fibres that compose the ventricles of the heart are so

accumulated and disposed, that their contraction originates, and communicates energetic impulse.

Fig. CLXXIV.—*View of the Outer Coats of the Smal. Intestine.*

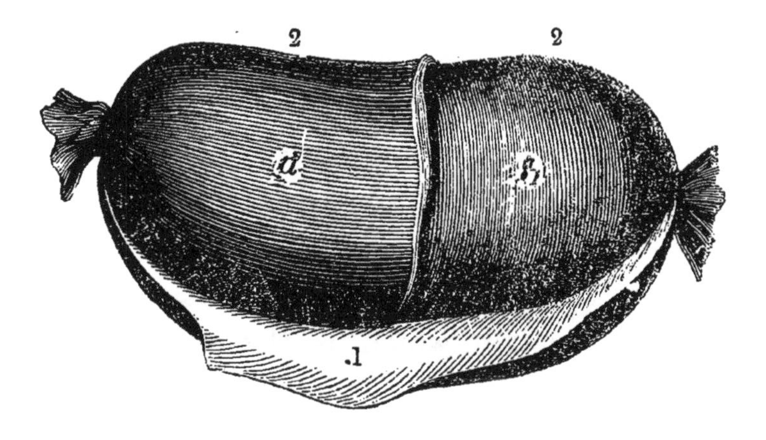

1. Peritoneal coat reflected off. 2. Muscular Coat consisting of—*a.* longitudinal fibres. *b.* Circular fibres.

The muscles of the arm are so accumulated and disposed that their contraction originates the like energetic impulse. Muscles so accumulated in the alimentary canal would have produced motion, indeed, but motion not only not accomplishing the end in view, but directly defeating it. In order to obtain the kind and degree of motion in this case required, the firm and thick muscle is attenuated into minute, delicate, and thready fibres, not concentrated in a bulky mass, so as to obtain by their accumulation a great degree of force; but spread out in such a manner as to form a thin and almost transparent coat. The tender fibres composing this delicate coat, by their contraction, produce two alternate, gentle, almost constant motions, called the peristaltic, from its resemblance to the motion of the earth-worm, and the antiperistaltic. By the peristaltic action motion is begun at once in several parts of the canal. Whenever the chyle is applied in a certain quantity to any part of the intestines, that part contracts, and makes a firm point, towards which the portions both above and below are drawn, by means of the longitudinal fibres which shorten the canal, and at the same time dilate the under part. By the antiperistaltic action, which is the exact reverse of the former, the chyle is turned over and over, and exposed to the orifices of the lacteal vessels; while, by the motion of the chyle forwards and backwards, and backwards and forwards, produced by these two actions constantly alternating with each other, its slow, gentle, but ultimately progressive course is secured.

682. The chyle thus gently moved along the extended surface of the jejunum and ilium, and still in its course acted upon in some degree by the secretions poured out upon the mucous membrane, successively disappears, until at the termination of the ilium (fig. CLXXI. 5) there is scarcely any portion of it to be perceived. It is taken up by the vessels termed lacteals.

683. The lacteal vessels (figs. 175 and 176), take their origin on the surface of the villi, by open mouths, too minute to be visible to the naked

eye, but distinguishable under the microscope. These minute, pellucid tubes, wholly countless in number, are composed of membranous coats so thin and transparent that the milky colour of their contents, from which they derive their name, is visible through them, and yet they are firm and strong. They present a jointed appearance (figs. CLXXVI. 4, and CLXXVII. 7). Each joint denotes the situation of the valves with which they are provided, and which are placed at regular distances along their entire course (fig. CXCII. 1 and 2). These valves, which are generally placed in pairs (fig. CXCII. 2), consist of a delicate fold of membrane of a semilunar form, one edge of which is fixed to the side of the vessel, while the other lies loose across its cavity (fig. CXCII. 2). So firm is this membrane, and so accurately does it perform the office of a valve, that even after death it is capable of supporting a column of mercury of considerable weight without giving way, and of preventing a retrograde course of the fluid. The lacteals are nourished by blood-vessels, and animated by nerves, and it is conceived that they must be provided with muscular fibres, or some analogous tissue, for they are obviously contractile, and it is by this contractile power that their contents are moved. The delicacy and transparency of the vessels, however, render it impossible to distinguish the different tissues which compose their walls.

684. If the mucous coat of the small intestines be examined some hours after a meal, the lacteals are seen turgid with chyle, covering its entire surface (fig. CLXXV. 1). These vessels, which are sometimes of such magnitude and in such numbers as entirely to conceal the ramifications of the blood-

Fig. CLXXV.

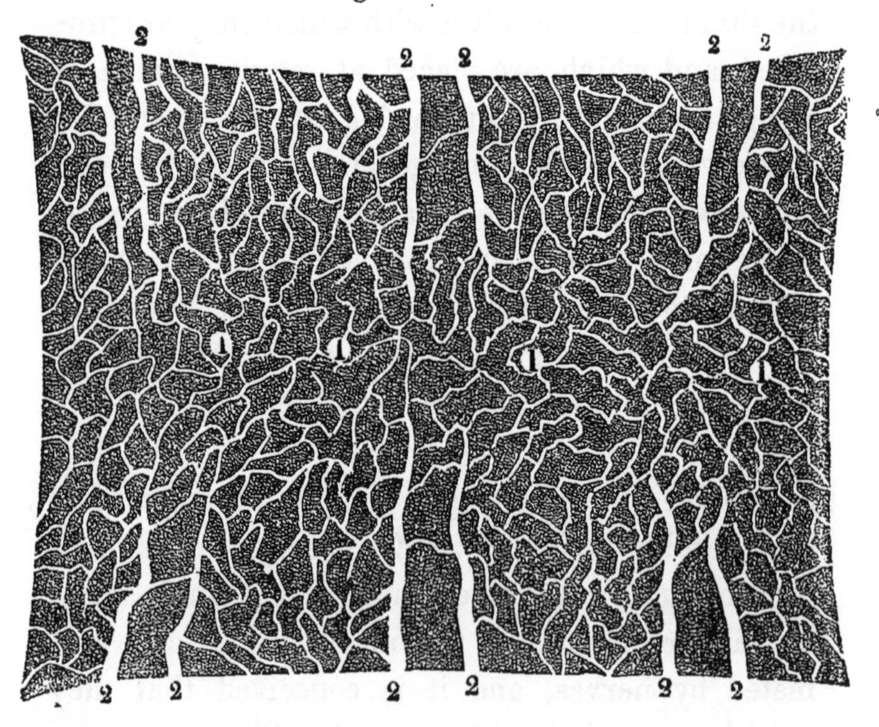

View of the inner surface of the ilium as it appears some hours after a meal. 1. The smaller branches of the lacteals, turgid with chyle, covering the surface of the intestine. 2. Larger branches of the lacteals formed by the union of the smaller branches.

vessels, unite freely with each other, and form a net-work, from the meshes of which proceed branches which, successively uniting, form branches of a larger size (fig. CLXXV. 2). These larger branches perforate the mucous coat and pass for some way between the mucous and the muscular

Fig. CLXXVI. *View of the course of the Lacteals.*

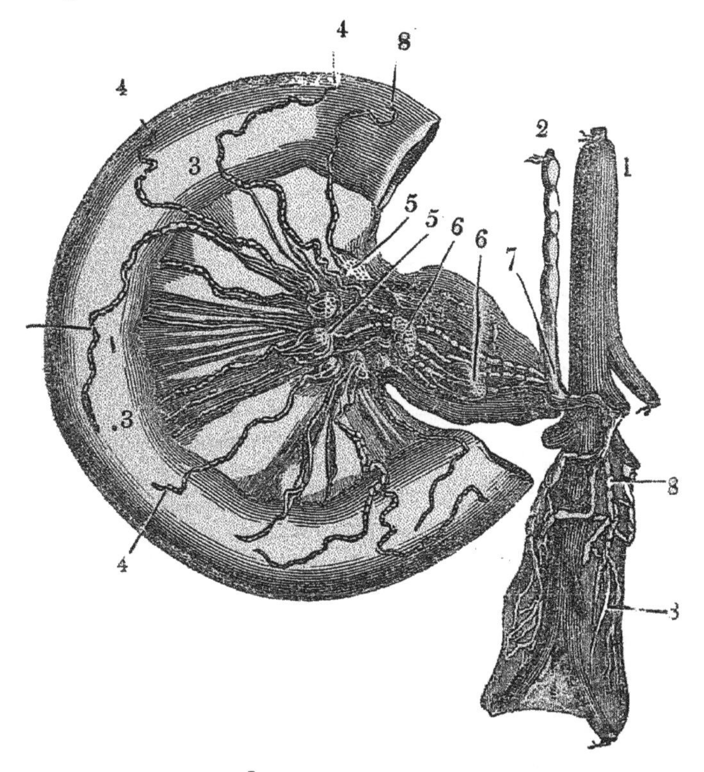

1. The aorta. 2. Thoracic duct. 3. External surface of a portion of small intestine. 4. Lacteals appearing on the external surface of the intestine after having perforated all its coats. 5. Mesenteric glands of the first order. 6. Mesenteric glands of the second order. 7. Receptacle for the chyle. 8. Lymphatic vessels terminating in the receptacle of the chyle, or commencement of the thoracic duct.

tunics: at length they perforate both the muscular and the peritoneal coats, when, from having been on the inside of the intestine, they get on the outside of it (fig. CLXXVI. 3, 4), and are included, like

Fig. CLXXVII.

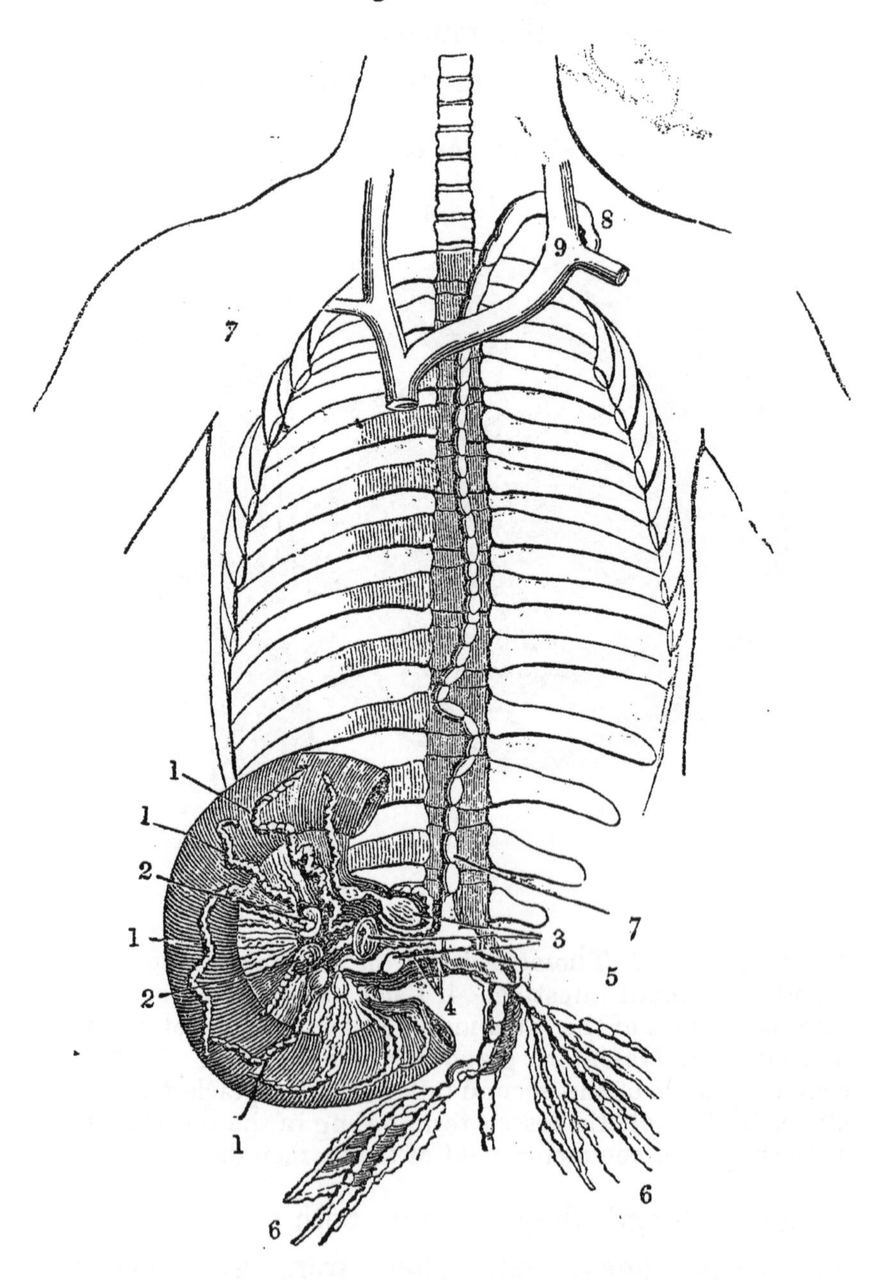

View of the course of the Thoracic Duct from its origin to its termination. 1. Lacteal vessels emerging from the mucous surface of the intestines. 2. First order of mesenteric glands. 3. Second order of mesenteric glands. 4. The great trunks of the lacteals emerging from the mesenteric glands, and pouring their contents into—5. The receptacle of the chyle. 6. The great trunks of the lymphatic or general absorbent system terminating in the receptacle of the chyle. 7. The thoracic duct. 8. Termination of the thoracic duct at—9. The angle formed by the union of the internal jugular vein with the subclavian vein.

the intestine itself, between the layers of the mesentery. All the different sets of lacteals converging and uniting together, form an exceedingly complicated plexus of vessels within the fold of the mesentery. Radiating from this plexus, the lacteals advance forwards until they reach the glands, called, from their being placed between the fold of the mesentery, the mesenteric (figs. CLXXVI. 5 and 6, and CLXXVII. 2 and 3); rounded, oval, pale-coloured bodies, consisting of two sets, arranged in a double row (figs. CLXXVI. 5 and 6, and CLXXVII. 2 and 3); the set nearest the intestine (fig. CLXXVII. 2) being considerably smaller than the succeeding set (fig. CLXXVII. 3).

685. On reaching the first series of glands (fig. CLXXVII. 2), the lacteals penetrate the substance of the gland, in the interior of which they communicate with each other so freely, and form such innumerable windings, that the gland seems to consist of a congeries of convoluted lacteals. Emerging from the first series of glands, the lacteals proceed on their course to the second series (fig. CLXXVII. 3), which they penetrate, and in the interior of which

they present the same convoluted appearance as in the first set. On passing out of this second series of glands, the lacteals unite together, and compose successively larger and larger branches, until at length they form two or three trunks (fig. CLXXVII. 4), which terminate in the small oval sac (fig. CLXXVII. 5), termed the receptacle of the chyle (receptaculum chyli).

686. In this oval sac or receptacle of the chyle (fig. CLXXVII. 5), which rests upon the second or the first lumbar vertebra, also terminate the trunks of the general absorbent vessels of the system (fig. CLXXVII. 6), called from the *lymph* or the pellucid fluid which they contain, lymphatics, as the lacteals are named from the lactitious or milky appearance of their contents.

687. The receptacle of the chyle produced forms the thoracic duct (fig. CLXXVII. 7), a canal about three lines in diameter. This tube rests upon the spinal column, ascends on the right side of the aorta, passes through the aortic opening in the diaphragm (fig. CXXXIV. 9, 10), and enters into the chest. Here it forms a transparent tube about the size of a crow-quill; it rests upon the bodies of the dorsal vertebræ; it continues to ascend still on the right side of the aorta, until it reaches the sixth or fifth dorsal vertebra, when changing its direction, it passes obliquely over to the left side (fig. CLXXVII. 7). From this point it continues its course upwards,

on the left side of the neck, as high as the sixth cervical vertebra; when suddenly turning forwards and a little downwards, it terminates its

Fig. CLXXVIII.—*Valve at the termination of the Thoracic Duct.*

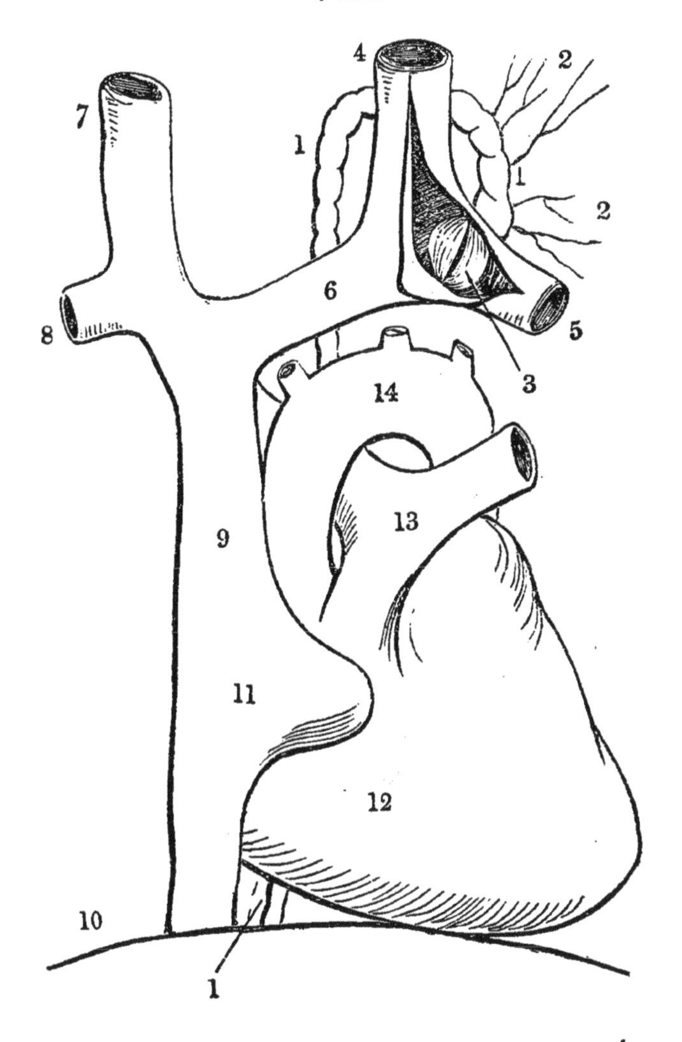

1. The Thoracic Duct. 2. Lymphatics entering the duct. 3. The vein laid open, showing the valve at the termination of the duct. 4. The left internal jugular vein. 5. The left subclavian vein. 6. The vein called innominata.

formed by the union of the internal jugular and subclavian veins. 7. The right jugular vein. 8. The right subclavian vein. 9. The superior cava formed by the union of the veins above. 10. The inferior cava formed by the union of the veins below. 11. The two venæ cavæ passing to the right auricle of the heart. 12. The heart. 13. The pulmonary artery dividing into right and left branches. 14. The aorta.

course in the angle formed by the union of the internal jugular with the subclavian vein (fig. CLXXVII. 8, 9). At its termination in these great venous trunks are placed two valves, which prevent alike the return of the chyle, and the entrance of the blood into the duct (fig. CLXXVIII.).

688. This account of the course of the thoracic duct is a description of the course of the chyle. Performing a double, circuitous, and slow circulation through the minute convoluted tubes of which the double series of mesenteric glands are composed, the chyle, in its receptaculum, is mixed with the contents of the lymphatic vessels, lymph (fig. CLXXVII. 6, 5), that is, organic matter brought from every surface and tissue of the body. Both fluids, chyle and lymph, mixed and mingled, flow together into the thoracic duct, by which in the course traced (687) they are poured into the blood, just as the venous torrent is rushing to the heart (fig. CLXXVIII. 6, 9, 11).

689. Thus, the final product of digestion, the chyle; particles of organized matter, the lymph; and venous blood, that is, blood which has already circulated through the system commingled, flow

together to the right heart, by which it is transmitted to the lungs, where all these different fluids are converted into one substance, arterial blood, to be by the left heart sent out to the system for its support.

690. While these processes are going on, another and a very important function is performed by the remaining portion of the alimentary canal. It is the office of this part of the apparatus to carry out of the body that portion of the aliment which is incapable of being converted into chyle. The preparation of the excrementitious part of the aliment for its expulsion constitutes the process of fecation. The organs in which this process is carried on, and by which the excrementitious matter, when duly prepared for its removal, is conveyed from the body, are the large intestines.

691. The large intestines (fig. CLXXIX.) consist of the cæcum, the colon and the rectum (fig. CLXXIX.). The cæcum varies in length from two inches to six; the colon is about five feet in length, and the rectum is about eight inches.

692. The ilium opens into the cæcum (fig. CLXXIX. 8, 10), just as the esophagus opens into the stomach. At this point the ilium is elongated, forming two concentric folds which join at their horns, and between the folds are placed a number of muscular fibres. In this manner is constructed a valve, which is termed the valve of the colon. It is placed in a transverse direction across the

Fig. CLXXIX.—*View of the Abdominal Portion of the Digestive Organs.*

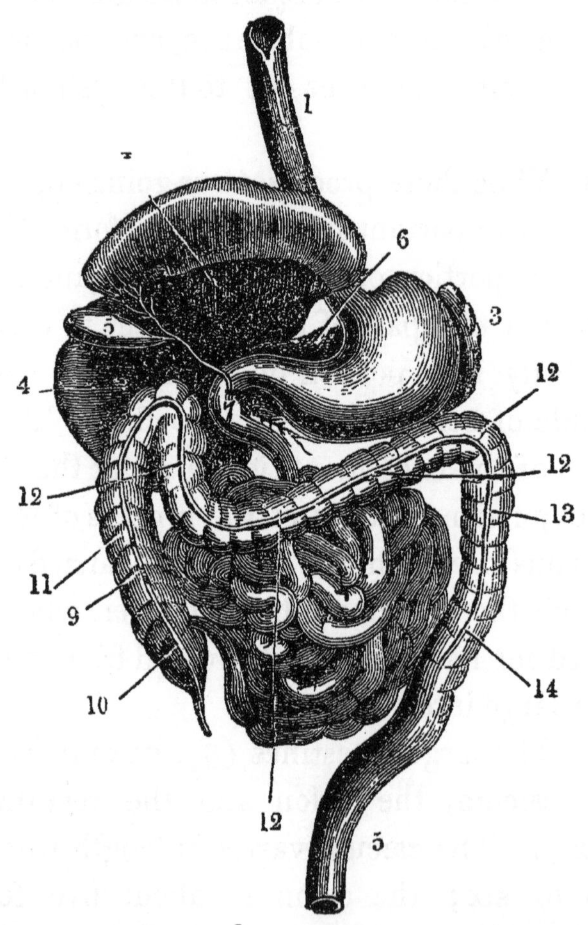

1. Esophagus. 2. Stomach. 3. Spleen. 4. Liver. 5. Gall-bladder with its ducts. 6. Pancreas with its duct. 7. Duodenum. 8. Small intestines. 9. Large intestines dividing into—10. Cæcum. 11. Ascending colon. 12. Arch of the colon. 13. Descending colon. 14. Signoid flexure here imperfectly represented. 15. Rectum.

intestine, and its action as a valve is very complete. It admits of the free passage of the con tents of the small intestines into the large, but it prevents the return of any portion of the contents of the latter into the former.

693. The colon is distinguished by its capacious size, its great length, and its longitudinal bands,

Fig. CLXXX.

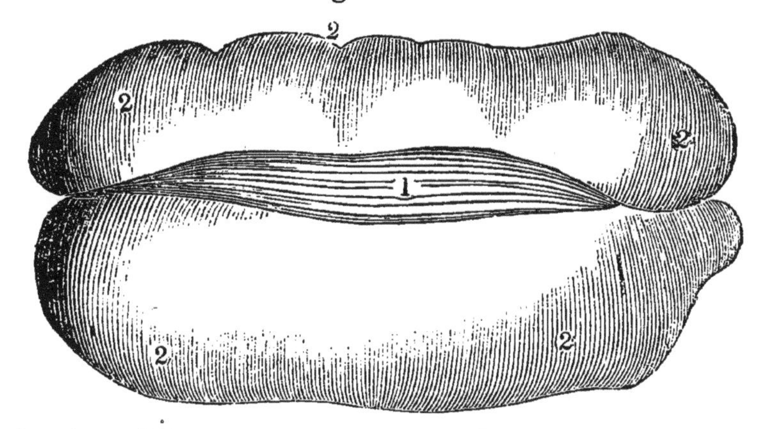

Portion of the large intestine, showing the arrangement of the muscular fibres. 1. The longitudinal fibres collected into bands, and forming larger fasciculi. 2. The circular fibres arranged as in the other intestines.

which consist of strong muscular fasciculi (fig. CLXXIX. 11). It is divided into an ascending portion which occupies the right iliac and hypochondriac regions (fig. CLXXIX. 11); the transverse portion, called its arch, which is placed directly across the epigastric region (fig. CLXXIX. 12), a descending portion which occupies the left hypochondriac region (fig. CLXXIX. 13), and a fourth portion, which being curved somewhat like the italic letter *S*, is called the sigmoid flexure, which occupies the left iliac region (fig. CLXXIX. 14). The sigmoid flexure terminates in the last portion of the alimentary canal, called the rectum (fig. CLXXIX. 15), which is placed in the hollow of the sacrum, and which follows the curvature of that bone (fig.

XLV. 5). The circular fibres of the rectum are accumulated at the termination of the bowel to form the internal sphincter of the anus. External to this is placed another set of fibres, which constitute the external sphincter.

694. The mucous membrane of the large intestines is disposed differently from that of the small intestines, and the mucous membrane of the colon still differently from that of the rectum. In the colon the mucous membrane, instead of being disposed in the form of valvulæ conniventes, is so arranged as to divide its whole surface into minute apartments or cells by which the descent of the fecal matter is retarded still more than the descent of the chyle by the valvulæ conniventes. Some particles of chyle do, however, continue to be separated from the fecal matter, even in the large intestines; and in order that nothing may be lost, a few valvulæ conniventes, with their lacteals, appear here also, while the cells of the colon, by retarding the descent of the fecal matter, allow time for the more complete separation and absorption of the chylous particles.

695. In the rectum the mucous membrane is plaited into large transverse folds, which disappear as the fecal matter descends into the bowel, accumulates in it, and distends it; an arrangement which gives to this portion of the intestine its power of distension, so closely connected with our convenience and comfort.

696. As soon as that portion of the alimentary matter which is transmitted to the large intestines reaches the colon it ceases to be alkaline, the distinctive character of the contents of the small intestines, and becomes acid, just as the whole alimentary mass is acid at the commencement of digestion in the stomach. It acquires albumen; its gases are no longer the same, for whereas pure hydrogen is contained in the small intestines, none is ever found in the large, but in the place of it, carbureted and sulphureted hydrogen; and now for the first time it receives its peculiar odour. As it continues to descend, its fluid parts are progressively absorbed, so that it becomes more and more solid, until it reaches the rectum, when it is almost dry. Here the accumulation of it goes on to a considerable extent, the peristaltic action at first excited by the distension of the rectum being, it would appear, counteracted by the contraction of the external sphincter of the anus. When, however, the distension of the bowel reaches a certain point, it produces a sensation which leads to the desire to expel its contents. The bowel is now thrown into action by an effort of the will, and that action is powerfully assisted by the descent of the diaphragm and the contraction of the abdominal muscles, actions also induced by an effort of the will. Thus the action of the first part of the digestive apparatus, that which is connected with the reception and partly with the deglution

of the food, is attended with consciousness, and is placed under the control of the will; the main portion of the digestive apparatus, that in which the essential part of the digestive process is carried on, is without consciousness, and is placed beyond the influence of volition; the last portion of the digestive apparatus, that connected with the expulsion of the non-nutrient portion of the aliment, again acquires sensibility and consciousness, and is placed under the control of the will. The striking differences in the arrangement of the muscular fibres in these different parts of the apparatus, in accordance with the widely different function performed by them; the powerful muscles connected with the prehension, mastication and deglutition of the food; the delicate and transparent tissue of fibres forming the muscular coat of the stomach and small intestines; the increase in the number and strength of the fibres of the large intestines, and the prodigious accession to them in the rectum, are adjustments not only exquisite and admirable in their own nature, but so indispensable to our well-being and comfort, that were the appropriate action of either to be suspended but for a short period, life would be extinguished, or if it could be protracted, it would be changed into a state of unbearable torment.

697. From the preceding account of the structure and action of the apparatus of digestion, on a comparison of all the phenomena, it appears that the

successive stages of the process are marked by the progressive approximation of the food to the nature of the blood. The main constituents, of the blood are albumen, fibrin, an oily principle, and red particles. Even in the chyme there are traces of albumen, with globules, not indeed to be compared in number with the red particles of the blood, smaller in size, and without colour, but still of an analogous nature. In the chyle of the duodenum the quantity of albumen is larger, there are traces of fibrin, and of an oily matter, and the number of the globules is increased. In the chyle, after its exit from the mesenteric glands, the albumen, the fibrin, the oil, the globules, and more especially the two first and the last, are greatly increased. But in the chyle when it reaches the thoracic duct, these principles are so augmented, concentrated, and approximated to the state in which they exist in the blood, that the chyle is now capable of undergoing the characteristic process of the blood; for as the blood, when drawn from a vein, undergoes spontaneous coagulation, so the chyle, when drawn from the thoracic duct, separates into three parts; a solid substance or clot, which remains at the bottom of the vessel; a fluid which surrounds the clot; and a thin layer of matter, which is spread over the surface of the fluid. The solid substance is analogous to the fibrin, and the fluid to the serum of the blood; while the layer of matter which is spread over the fluid is of an oily nature: moreover, the chyle, when in contact with the air, quickly changes to a red colour, and abounds with minute particles of various sizes, but the largest of which is not yet equal to the diameter of the red particles of the blood.

698. The changes wrought upon the food, by which it is thus approximated to the chemical composition of the blood, are effected, as has been shown, partly by the gastric and intestinal juices, and partly by matters combined with the food highly animalized in their own nature, and endowed with assimilative properties, as the salivary secretion mixed with the food during mastication; the pancreatic and biliary secretions mixed with the food during the conversion of the chyme into chyle; and the mesenteric secretions mixed with the elaborated chyle of the mesenteric glands, and lastly, organized particles which have already formed a part of the living structures of the body mixed with the chyle under the form of lymph in the thoracic duct.

699. The lymph, until lately regarded as excrementitious, is really highly animalized, partly combined with the chyle as its last and highest assimilative matter; whence the compound formed by the admixture of chyle and lymph is far more proximate to the blood than the purest and most concentrated chyle; and partly returning with the chyle to the lungs, to receive there a second depuration, and thereby a higher elaboration.

700. There is evidence that there is a series of organs specially provided for the elaboration of the lymph no less than of the chyle. There are organs manifestly connected with the digestive apparatus, to which physiologists have found it extremely difficult to assign a specific office. These organs have a structure in some essential points alike; that structure is strikingly analogous to the organization of glands: like glands, they receive a prodigious quantity of arterial blood, and are supplied with a proportionate number of organic nerves; yet they are without an excretory duct. The organs in question are the bodies called the renal capsules, placed above the kidneys; the thyroid and thymus glands situated in the neck, and the spleen in close connexion with the stomach.

701. These organs, however analogous in structure to glands, cannot, it has been argued, be secreting organs, because they are destitute of an excretory duct, do not manifestly form from the blood any peculiar secretion, or, if they do, since there are no means of detecting where it is conveyed, it is impossible to understand how it is appropriated. But if these organs collect, concentrate, and elaborate lymph, preparatory to its admixture with the chyle and to its being sent a second time into the blood to undergo a second process of depuration, they perform the function of glands; and their want of an excretory duct, which has hitherto rendered their office so obscure, is accounted

for; they do not need distinct tubes for the transmission of any product of secretion; the lymphatic vessels which proceed from them and which convey the fluid they elaborate into the receptacle of the chyle, are their excretory ducts. That one of these organs, the spleen, is specially connected with the elaboration of the lymph, is manifest, both from its chemical nature and from the remarkable change which takes place in the chyle the moment the lymph from the spleen is mixed with it. Tiedemann and Gmelin state, as the uniform result of their observations and experiments, that the quantity of fibrin contained in the chyle is greatly increased, and that it actually acquires red particles as soon as the lymph from the spleen is mixed with it, and that the lymph from the spleen superabounds both with fibrin and with red particles. That the organs just enumerated, with the spleen, perform a similar function, is inferred from their being, like it, of a glandular structure, and without any excretory duct. If the spleen be really one of a circle of organs appropriated to a function such as is here supposed, a purpose is assigned to it adequate to its rank in the scale of organization; inferior to few, if its importance be estimated by the quantity of arterial blood with which it is supplied; yet this is the organ for which Paley could find no better use than that of serving for package.

702. But in whatever mode the lymph be elaborated, it is certain that it consists of matter highly animalized, and that its most important principles, its albumen, its fibrin, its globules, and even its salts, are in a chemical condition closely resembling that in which they exist in the blood.

703. It will appear hereafter that all the proximate principles of which the body is composed are reducible by analysis to three, namely, sugar, oil, and albumen: of these, sugar and oil are the least, and albumen the most highly organized. Every alimentary substance must contain at least one of these proximate principles, and in the various articles which compose an ordinary meal always two, and often all three, are afforded in abundance. From the phenomena which have been stated, it is clear that the digestive organs, in acting on these principles, exert the following powers.

1. A solvent power. The first action of the stomach on the alimentary substances presented to it is to reduce them to a fluid state. No substance is nutritious which is not a fluid, or capable of being reduced to a fluid. The stomach reduces alimentary substances to a fluid state by combining them with water. Water enters into the composition of organized bodies in two states, as an essential and as an accidental element. A quantity of water is contained in sugar when reduced to its dryest state; this water cannot be dissipated without the decomposition of the sugar;

it is therefore an essential constituent of the compound. Water is combined with sugar in its moist state: of this water much may be removed without destroying the essential properties of the sugar: this part of the water is therefore said to be an accidental constituent of the sugar. In most cases organized bodies contain water in both these forms; and though it is commonly impossible to discriminate between the water that is essential and that which is accidental, yet the mode of union among the elements of bodies in these two states of their combination with water are essentially different. The stomach has the power of combining water with alimentary substances in both these forms. Thus fluid albumen, or white of egg, presented to the stomach is immediately coagulated or converted into a solid. Soon this solid begins to be softened, and the softening goes on until it is again reduced to a fluid. What was fluid albumen in the white of egg is now fluid albumen in chyme; but the albumen has undergone a remarkable change. Out of the stomach the albumen of the egg may be converted by heat into a firm solid; but the albumen of the chyme is capable of being converted only into a loose and tender solid In passing from its state in the egg to its state in the chyme, the albumen has combined with a portion of water which has entered as an essential ingredient into its composition. By this combination the compound is reduced from what may be called a strong to a weak state. This is the first action exerted by the stomach on most alimentary substances. They are changed from a concentrated to a diluted, from a strong to a weak state: the power by which the stomach effects this change is called its reducing power, and the agent by which it accomplishes it is the gastric juice; the essential ingredient of which has been shown to be muriatic acid, or chlorine (639, *et seq.*). The muriatic acid obtained from the common salt of the blood is poured in the form of gastric juice into the stomach, dissolves the food, combines it with water, reduces it from a concentrated solid to a dilute fluid; and thus brings it into the condition proper for the subsequent part of the process.

2. A converting power. Since whatever be the varieties of food, the chyme invariably forms a homogeneous fluid, the stomach must be endowed with the power of transforming the simple alimentary principles into one another; the saccharine into the oily, and the oily into the albuminous. The transformation of the saccharine into the oleaginous principle is traceable out of the body in the conversion of sugar into alcohol, which is essentially an oil. That the same transformation takes place within the body is indubitable. The oleagenous and the albuminous principles are already so nearly allied in nature to animal substance that they do not need to undergo any essential change in their composition.

3. A completing power. When the alimentary substances have been reduced and formed into chyme, when the chyme has been converted into chyle, and when the chyle absorbed by the lacteals is transmitted to the mesenteric glands, it undergoes during its passage through these organs a process the direct reverse of that to which it is subjected in the stomach; for whereas it is the office of the stomach to combine the alimentary substances with water, it is one office of the mesenteric glands to remove the superfluous water of the chyle; to abstract whatever particles of matter may be contained in the compound which are not indispensable to it, and to concentrate its essential constituents; and consequently these organs exert on the digested aliment a completing, in contradistinction to a reducing power.

4. A vitalizing power. When sugar is converted into oil, when oil is converted into albumen, when albumen, by the successive processes to which it is subjected is completed, that is, when the alimentary substances are made to approximate in the closest possible degree to the nature of animal substance, they must undergo a still further change, more wonderful than any of the preceding, and far more inscrutible; they must be endowed with vitality; must be changed from dead into living matter. Living substance only is capable of forming a constituent part of living substance. The ultimate action of the

digestive organs is the communication of life to the food, to which last and crowning process the reducing, converting, and completing processes are merely subordinate and preparatory. Of the agency by which this process is effected we are wholly ignorant; we know that it goes on; but the mode in which it is accomplished is veiled in inscrutable darkness.

704. Blood is alive; blood is formed from the food; life is communicated to the food before it is mixed with the blood. The blood is essentially albumen, which it contains in the form of albumen properly so called, in that of fibrin, and in that of red particles. In the thoracic duct the strong albumen of the lymph is mixed with the weaker albumen of the chyle. At the point where the thoracic duct terminates in the venous system, lymph and chyle are mixed with venous blood, and all commingled are borne directly to the lungs. There the carbon with which the venous blood is loaded is expelled in the form of carbonic acid gas; the particles of the lymph undergo some, as yet, unknown change, exalting their organization; and the water hitherto held in chemical union with the weak albumen of the chyle, is separated and carried out of the system together with the carbonic acid gas in the form of aqueous vapour. By this removal of its aqueous particles the ultimate completion is given to the digested aliment; and the weak and delicate albumen of the chyle is converted into the strong and firm albumen of the blood.

705. It has been stated (539), that though gelatin enters abundantly into the composition of many tissues of the body, and performs most important uses in the economy, it is never found in the blood; that it is formed from the albumen of the blood by a reducing process, in consequence of which carbon is evolved, which unites with the free oxygen of the blood, forming carbonic acid, thus conducing, among other purposes, to the production of animal heat. It is equally remarkable, that though the lymphatics or absorbents arise in countless numbers from every tissue of the body, and are endowed with the power of taking up every constituent particle of every organ, solid as well as fluid, yet gelatin is never found in the lymphatic vessels. The lymphatics contain only albumen in a form far more proximate to the blood than that of the chyle; consequently, before the gelatin of the body is taken up by the lymphatics, it must be reconverted into albumen; that is, the absorbed gelatin must undergo a process analogous to that which gelatin and other matters undergo in the stomach and duodenum; it follows that the digestive process is not confined to the stomach and duodenum, but is carried on at every point of the body. Hence there are two processes of digestion, a crude and a refined process. The crude process is carried on in the stomach and duodenum, in which dead animal matter is converted into living substance, as yet, however, possessing only the lowest kind of vitality.

The capillary arteries receiving the substance thus prepared for them, build it up into structure perhaps the lowest and coarsest, the least organized, and capable of performing only the inferior functions.

706. Capillary arteries in countless numbers terminate in the tissues in membraneless canals (304 and 310). Particles of the blood are seen to quit the arterial stream and to enter into the tissues, becoming a component part of them: other particles are seen to quit the tissues and to enter the current of the blood. The latter are probably organic particles, to which a certain degree of elaboration has been already given, now transmitted to the capillary veins, to be carried back to the lungs to undergo there a further depuration, fitting them on their return to the system for a higher organization.

707. Thus the lymphatic vessels, analogous in so many other respects to the veins, are probably similar to them in this also—that they take up from the tissues particles already organized, in order to submit them to processes which communicate to them a progressively higher organization. The notion that the contents of the lymphatics consist of worn-out particles, capable of accomplishing no futher purpose in the economy, is not tenable:—

1. Because it is not analogous to the ordinary operations of nature to mix wholly excrementitious matter with a substance for the production, ela-

boration, and perfection of which, she has constructed such an expensive apparatus.

2. Because, on the other hand, the admixture of matter already highly animalized with matter, as yet but imperfectly animalized, exalts the nature of the latter, and is conducive to its complete animalization.

3. Because the lymph, almost wholly albuminous, is already closely allied in nature to the blood; it is, therefore, reasonable to infer, that it is matter passing through an advancing stage of purification and exaltation.

4. Because this plan of progressive organization is in harmony with the ordinary operations of nature, in which there is traceable a successive ascent from the low to the high, the former being preparatory and necessary to the latter. The tender and delicate organs of animal life, the brain, the nerves, the apparatus of sense, the muscles, inasmuch as they perform the highest functions, probably require to be constructed of a more highly organized material, for the production of which the matter primarily derived from crude aliment is subjected to different processes, rising one above the other in delicacy and refinement; by each of which it is made successively more and more perfect, until it acquires the highest qualities of living substance, and is capable of becoming the instrument of performing its most exalted functions.

CHAPTER XI.

OF SECRETION.

Nature of the function—Why involved in obscurity—Basis of the apparatus consists of membrane—Arrangement of membrane into elementary secreting bodies—Cryptæ, follicles, cæca and tubuli—Primary combinations of elementary bodies to form compound organs—Relation of the primary secreting organs to the blood-vessels and nerves—Glands simple and compound—Their structure and office—Development of glands from their simplest form in the lowest animals to their most complex form in the highest animals—Development in the embryo—Number and distribution of the secreting organs—How secreting organs act upon the blood—Degree in which the products of secretion agree with, and differ from, the blood—Modes in which modifications of the secreting apparatus influence the products of secretion—Vital agent by which the function is controlled—Physical agent by which it is effected.

708. SECRETION is the function by which a substance, gaseous, liquid, or solid, is separated or formed from the nutritive fluid. It is a function as necessary to the plant as to the animal, and indispensable alike to the life of both. It is of equal importance

to the preservation of the individual and to the perpetuation of the species. In all living beings secretions are separated from the nutritive fluid, and added to the aliment to assist in converting it into nutriment, and are separated from the nutriment to maintain the composition of the nutritive mass in a state fit for the continued performance of the act of nutrition, and to form the germ on the development of which the continuance of the species depends.

709. The secretions of the plant, varied and abundant, are indispensable to its nourishment, growth, and fructification. The secretions of the animal more diversified, and far more constantly performed, increase in number and elaborateness in proportion to the range and intensity of the vital endowments and actions. In all animals high in the scale of organization, and especially in man, the products of secretion are vast in number, and exceedingly complex in nature,—membrane, muscle, brain, bone;—the skin, the fat. the nail, the hair;—water, milk, bile, wax, saliva, gastric juice;—whatever substances enter as constituents into the corporeal structure;—whatever substances are specially produced, in order to perform some definite purpose in the economy;—whatever substances are separated from the mass, and carried out of the system on account of their useless or noxious properties:—all are derived from the nutritive fluid,

the blood, and are formed from it by the process of secretion.

710. In this function are included the most secret and subtle processes of the vital economy,—the ultimate actions of the organic life. Of the real nature of those actions nothing definite is known; and they are modified by agencies over which the art and skill of the experimentalist can exert no adequate control. It is not wonderful therefore that they should be involved in obscurity: nevertheless, when all the phenomena are collected and compared, much of the mysteriousness in which the function appears at first view to be involved vanishes.

711. The apparatus of secretion is infinity varied in form: when examined in its complex combinations it appears inextricable in structure, but the diligence and skill of modern research have unfolded much of its mechanism, and enabled us to trace the successive steps by which it passes from its simple to its complex condition.

712. To form an organ of secretion there must be an artery, a vein, a nerve, an absorbent, and a sufficient quantity of cellular tissue to allow of the free expansion of these vessels and of their complete intercommunication. Membrane constitutes such an organ; for membrane is composed of arteries, veins, nerves, and absorbents sustained and connected by cellular tissue. Hence membrane con-

stitutes a secreting organ, in its simplest form. The most important secreting membranes are the serous (30), the cutaneous (34), and the mucous (33).

713. Serous membrane which lines the great cavities of the body, and which gives an external covering to the organs contained in them (fig. LX. a, c), forms an extensive secreting surface. Synovial membrane, or that which covers the internal surface of joints, and which constitutes an important portion of the apparatus of locomotion, is essentially the same in structure and office.

714. Cutaneous membrane, or the skin, which forms the external covering of the body, is an organ in which manifold secretions are constantly elaborated; but the skin is only a modification of the membrane which lines the interior of the body, the mucous. Mucous membrane forms the basis of the secreting apparatus placed in the mouth, fauces, esophagus, stomach, and intestines in their whole extent; of the secreting apparatus auxiliary to that of the alimentary canal, namely, the pancreas and the liver; probably also of the mesenteric, or lacteal glands, together with the vast system of lymphatic glands, and certainly of the glands of the larynx, trachea, bronchi and air vesicles of the lungs. Hence, while membrane forms the basis of the secreting apparatus in general, mucous membrane is far more extensively

employed in its construction than any other form of membrane.

715. 1. In the construction of the secreting apparatus, membrane disposed in the simplest form, constitutes merely a uniform, smooth, extended surface. Serous membrane is always disposed in this simple mode. The costal pleura which lines the internal surface of the walls of the chest (fig. LX. a); the pulmonary pleura which is continued from the walls of the chest over the lungs (fig. LX. 5); the peritoneum which lines the internal surface of the cavity of the abdomen, and which is reflected over the viscera contained in it (fig. LX. c, and 6, 7, 8, &c.); the synovial membrane which covers all the articular surfaces; the arachnoid membrane which envelopes the brain, form simple continuous, serous, secreting surfaces. On the contrary, mucous membrane is never dis-

Fig. CLXXXI.

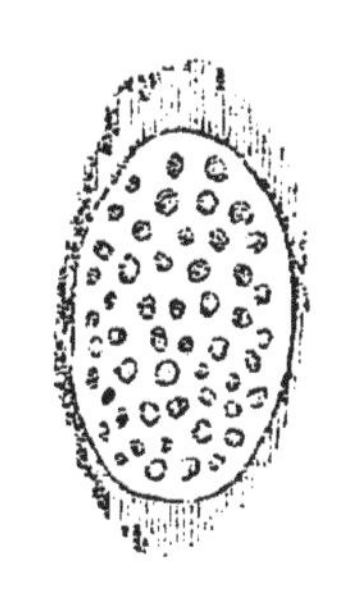

A portion of the mucous surface of the intestines, showing some of the mucous glands which present the appearance of fovæ or cryptæ.

posed in this perfectly simple mode; even when it forms a continuous surface, as in the lining, which it affords to the alimentary canals, it is more or less plaited into folds or rugæ (fig. CLXVII. 1).

716. 2. The second disposition of membrane in the construction of the secreting apparatus, is the depression of it into a minute pit or fova, called a crypt (CLXXXI.), which is sometimes inclosed on all sides, forming a cell or vesicle (fig. CXXXVIII.).

717. 3. Next, the vesicle, instead of being

Fig. CLXXXII.

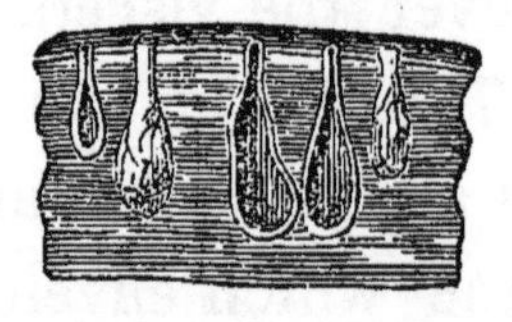

Portion of the skin and cellular tissue, showing the sebaceous follicles, as seen under the microscope very highly magnified. 1. The external surface of the follicles with the blood-vessels ramifying upon it. 2. Follicles laid open, showing the interior cavity into which the secreted fluid is poured.

rounded, is elongated into a peduncle or neck, not unlike the neck of a bottle (fig. CLXXXII. 1). This pedunculated vesicle is called a follicle.

718. 4. Then, the follicle is somewhat elongated, without neck and without terminal expansion (fig. CLXXXVI. 1); and this is called a cæcum or pouch.

719. 5. And, lastly, the cæcum itself is elon-

gated; so that instead of presenting the appearance of a pouch, it rather resembles a tube (fig. CLXXXV. 1), and is accordingly named tubulum.

720. In the construction of the secreting apparatus, membrane, then, may be said to be disposed into four elementary forms constituting cryptæ or vesicles, follicles, cæca and tubuli. Membrane, disposed into these elementary forms, constitutes the simple bodies by the accumulation and the varied arrangement of which the compound organs are composed. There is no other known element which enters into the composition of the most complex secreting organ.

721. One of these elementary bodies may exist as a simple organ, or many may be collected into a mass to form a compound organ. When single they are called solitary: when collected into a mass, aggregated. Each elementary body has a mode of aggregation peculiar to itself. Vesicles aggregate by clustering together (fig. CXXXVIII.), and adhering as if by a common stem (fig. CXXXVIII.); follicles by uniting at their orifices (fig. CLXXXIII.), and forming masses which are disposed either in a linear direction (fig. CLXXXIII.) or in fasciculi (fig. CLXXXIV.); cæca by forming bundles, parallel or branched (fig CLXXXVI.); and tubuli by forming masses straight (fig. CLXXXV.), tortuous or convoluted (figs. CLXXXV. and CLXXXIX.)

722. When a single elementary body, as a vesicle

Fig. CLXXXIII.

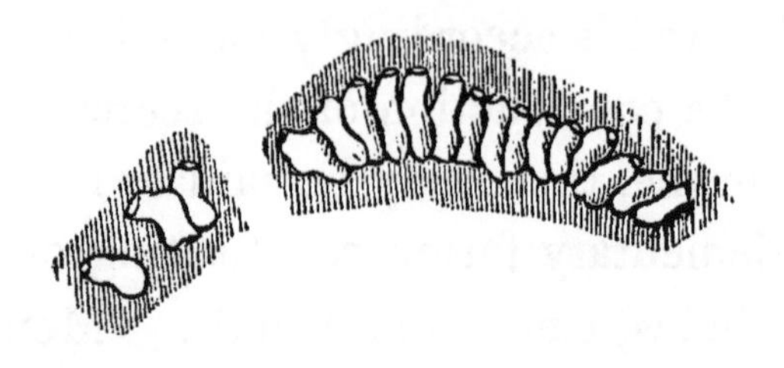

Aggregated follicles disposed in a linear direction, here represented of their natural size, as seen near the mouth in the goose.

or follicle, forms a distinct secreting organ, the matter secreted is elaborated at the inner surface of

Fig. CLXXXIV.

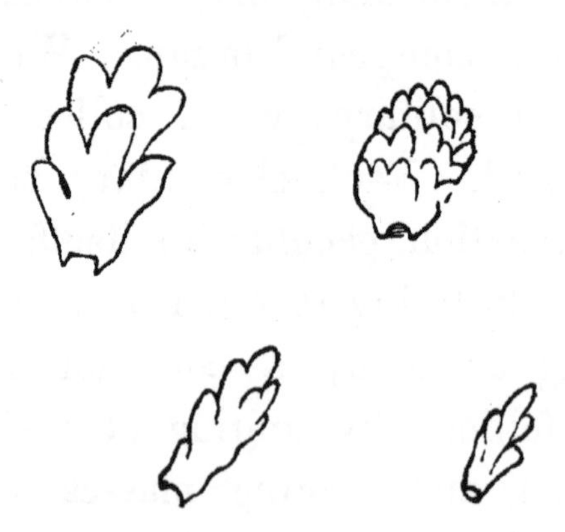

Conglomerated follicles.

the organ (fig. CLXXXII. 2), and is contained within its cavity. When needed it quits this cavity through the walls of the vesicle, or at the orifice of the follicle, on the application of the appropriate stimulus. When a number of cryptæ or vesicles are aggregated into clusters, the individual vesicles sometimes open by distinct orifices into a common receptacle or sac (fig CLXXXIV.). When follicles are aggregated into a mass, and the mass is disposed in a linear direction (fig. CLXXXIII), each follicle pours out its secreted matter by its own orifice (fig. CLXXXIII.) ; but if conglomerated, into a common mass by a common orifice (fig. CLXXXIV.).

723 In like manner, in some very simple arrangements of cæca and tubuli, each body opens by its own distinct orifice (fig. CLXXXV. 2). But in the

Fig. CLXXXV.

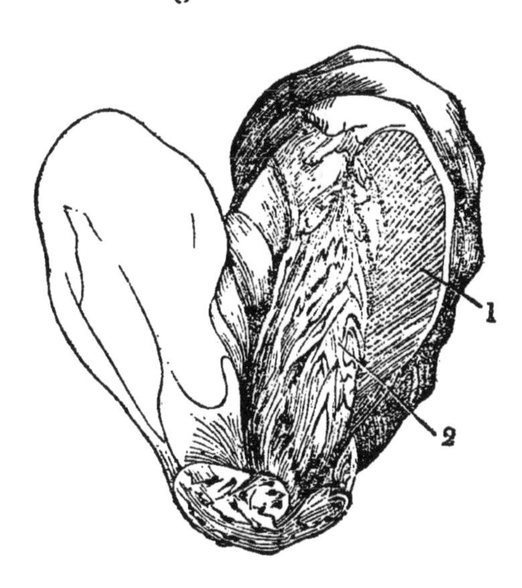

1. Parallel tubuli, opening by distinct orifices into—2. A common cavity.

more complex arrangements of these bodies; it is indispensably necessary to modify this mode of parting with their contents. When the elementary bodies are aggregated into dense, thick masses (fig. CLXXXIX.), when layer after layer of these masses containing myriads of myriads of follicles, cæca, or tubuli, are superimposed one upon another, (fig. CLXXXIX.), it is impossible that each individual

Fig. CLXXXVI.

Branched cæca, showing—1. The cæca terminating in—2. Excretory ducts which unite to form—3. A common trunk.

body can have a separate orifice. In this case a minute tube springs from each body (fig. CLXXXVI. 2); and a complete connexion is established between all the individuals composing the mass by the free inter-communication of these tubes (fig. CLXXXVI. 2). Of these tubes the minutest unite together, and form larger branches (fig. CLXXXVI. 2); these larger branches again uniting form still larger branches (fig. CLXXXVI. 2), until, by their successive union, the branches form at length a single trunk (fig. CLXXXVI. 3), with which all the individual branches, whether great or small, communicate, and into which they all pour their contents (fig. CLXXXII. 2, 3). The bodies from which these tubes take their origin, and the minute tubes themselves, are called secreting canals (fig. CLXXXII. 1, 2); the common trunk formed by their union is termed the excretory duct (fig. CLXXXII. 3). The secreting canals contain the secreted matter; the excretory duct collects this matter, and conveys it to the part of the body in which it is appropriated to the specific purpose which it serves in the economy.

724. The basis of the secreting canals consists, then, of membrane disposed in one or other of the elementary forms described (712, *et seq.*). These secreting canals constitute a peculiar system of organs wholly different from all the other organs of the body. The form of these organs, their structure and their relation to the blood-vessels and nerves, have formed subjects of laborious investigation and

of keen controversy during several centuries. The honour of discovering the exact truth on these points is due to very recent researches.

725. Malpighi, an Italian, who flourished at Bologna in the middle of the 17th century, was the first to establish a special inquiry into the intimate structure of the secreting apparatus. After many years of laborious examination he arrived at the conclusion that a minute sac or follicle is invariably interposed between the termination of the capillary artery and the commencement of the excretory duct. According to him, the capillary artery conveys the blood to the follicle, separates from the blood the substance secreted, and the excretory duct arising from one extremity of the follicle conveys the secreted fluid, when duly prepared, to its destined situation. By injection, by dissection, by the microscope, by experiment on living animals, and by the phenomena of disease, he conceived that he had demonstrated that this is the true structure of the secreting apparatus in its most complex form. This view was generally acquiesced in by his contemporaries and by succeeding anatomists and physiologists; and in the time when Ruysh wrote was the received opinion.

726. Ruysh, who flourished at Amsterdam, and was contemporary with Malpighi, but a younger man, and who published on the glands about twenty years after Malpighi, according to the account of Haller, " employed wonderful pa-

tience, with the assistance of his daughters, in rendering all his preparations elegant and beautiful, being equally skilled in the methods of softening, hardening, filling, and drying." Of Ruysh it was said that while others, in their anatomical preparations, merely exhibited the horrid features of death, he preserved the human body in all the freshness of life, even to the expression of the features. The fineness of his injections, the dexterity with which he unfolded the minute vessels, nerves, and absorbents, and exhibited their combinations and relations in the most delicate structures, the skill with which he preserved his preparations in transparent fluids, and the elegance with which he displayed them in their natural forms and folds, excited universal admiration; and philosophers, statesmen, princes, kings, all the learned and noble of the day, crowded to his museum.

727. By his superior method of injecting, Ruysh conceived that he was able completely to disprove Malpighi's doctrine. He maintained that the bodies which Malpighi mistook for sacs or follicles are in reality convoluted vessels; that these vessels are capable of being completely unravelled; that, when unfolded, their continuity with the excretory duct is perfectly demonstrated; that secretion is performed by the capillary artery itself, without the intervention of any other organ; and that when the secreted substance is duly

prepared, it is poured by the capillary directly into the excretory duct.

728. Modern research has demonstrated that the opinion of Malpighi approaches nearer the truth than that of Ruysh, who appears to have mistaken the secreting canals for the ultimate division of the arterial vessels. Malpighi, indeed, did not succeed in discovering the elementary bodies of which the secreting apparatus is composed; but he arrived at the very verge of the truth. Profiting by the art which Ruysh brought to so much perfection, by the facts which Malpighi disclosed, and, above all, by the improved structure of the microscope, and the increased skill which has been acquired in the manipulation of the instrument, the modern physiologist is enabled to see what was formerly beyond the cognizance of sense, and to demonstrate what before could only be matter of conjecture. Availing himself of these advantages with consummate skill, and applying himself to the task with indefatigable industry, Professor Müller, of Berlin, has investigated the structure of the secreting apparatus in the whole animal kingdom, and has traced the progressive development of the several secreting organs through the entire animal series, from their simplest form in the lowest animal, to their most complex in the highest.

729. From the researches of this physiologist, and from the labours of others, his countrymen and

contemporaries, who have engaged in the investigation with an ardour second only to his own, it is demonstrated that the secreting apparatus of the animal body is disposed in one or other of the elementary forms which have been described. The blood-vessels are distributed upon the walls of these elementary bodies, whether simple cryptæ follicles, cæca, or tubuli, or whether these bodies are accumulated and combined into the largest and most complex series of secreting canals, just as the branches of the pulmonary artery are distributed upon the walls of the air-vesicles in the rete mirabile of the lungs. The air-vesicles of the lungs are secreting organs, and afford an excellent example of the mode in which the blood-vessels are distributed upon the walls of the elementary secreting bodies. The arteries do not form continuous tubes with the secreting bodies or their excretory ducts, as was maintained by Ruysh; neither is the secreting body interposed between the termination of the artery and the commencement of the excretory duct, as was thought by Malpighi; but the ultimate divisions of the arteries are spread out upon the walls of the secreting bodies, where they terminate in veins by a delicate vascular network (fig. CLXXXVII. 2). The minutest branch of the artery is always smaller than the minutest secreting body on the walls of which it is distributed. According to Müller, the arteries, spread out upon the walls of the secreting bodies, form a distinct and peculiar system of vessels visible under the microscope. In the more complex secreting organs, immediately before reaching their distribution upon the walls of the secreting canals, the ultimate divisions of the arteries form an intricate and delicate net-work (fig. CLXXXVII. 2). When at length they reach the secreting canals the arteries no longer divide and subdivide, but are always of the same uniform size in the same secreting organ, though their magnitude is different in every different kind of secreting organ. These ultimate divisions of the arteries are the proper

Fig. CLXXXVII.

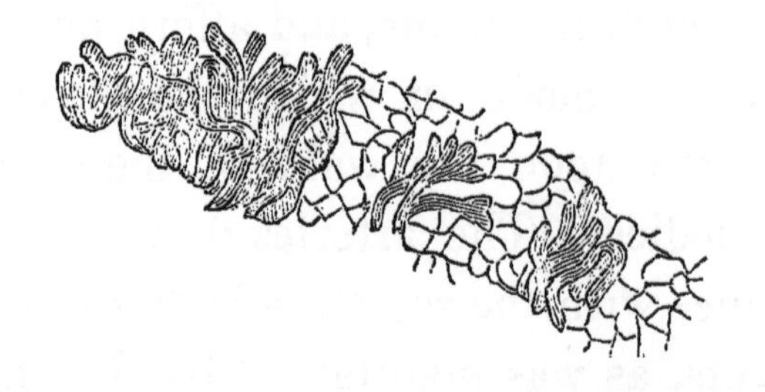

A thin portion of the surface of the kidney taken from the scianus, showing—1. The termination of the cæca forming the uriniferous duct; and—2. A delicate vascular net-work, consisting of capillary blood-vessels about to be distributed on the walls of the cæca.

capillary arteries. It is in these arteries that the changes are wrought upon the blood which it is the object of the various processes of secretion to effect. In the walls of these arteries there are visible no pores, no apertures, no open extremities by which the secreted fluid, when formed from the blood, is conveyed into the cavity of the secreting canals; it probably passes through the walls of the vessels into the secreting canals by the process of endosmose (804).

730. Secreting organs are very abundantly supplied with nerves, which are derived for the most part from the organic portion of the nervous system; although for the reasons assigned (vol. i. p. 77, *et seq.*) sentient nerves are mixed with the organic. The more important secreting organs have each a distinct net-work or plexus of organic nerves, which surround the blood-vessels distributed to the organ, (fig. CLXX. 3), and which envelopes more especially the arterial trunks and their larger branches (fig. CLXX. 3). From these plexuses nervous filaments spring in countless numbers (fig. CLXX. 3), which are spread out upon the walls of the arteries, just as the arteries are spread out upon the walls of the secreting canals. The nerves never quit the arteries; are never spent upon the membranous matter which forms the basis of the secreting organ, but are lost upon the walls of the capillary arteries. The nerves uniformly increase in number and size as the arteries diminish in magnitude and as their capillary terminations become thinner and thinner.

731. When the secreting apparatus consists of simply extended membrane, a close net-work of capillary arteries with their accompanying nerves is spread out over the whole extent of the secreting surface. This simple arrangement is sufficient to separate from the blood the simple secretion in this case required.

732. When the secreting apparatus consists of simple cryptæ, follicles, cæca. or tubuli, a similar net-work of capillary arteries and nerves is spread out on the sides of this more extended surface. The more elaborate secretion now formed is received into the interior of these organs, where it remains for some time, and whence it is ultimately conveyed as it is needed by the actions of the system.

733. But when the secreting apparatus consists of aggregates of cryptæ, follicles, cæca, and tubuli, with their net-works of arteries and nerves, a much more complex structure is built up, which is destined to perform a proportionably elaborate function. An aggregation of these secreting bodies into a large mass, enveloped in a common membrane, so as to form a distinct body of a solid consistence, constitutes the organ termed a gland. Simply extended membrane, with its apparatus of arteries and nerves does not constitute a gland. Simple cryptæ, follicles, cæca, and tubuli, with their larger apparatus of arteries and nerves, do not constitute a gland. The first is simply secreting surface; the second are simply secreting cryptæ, follicles, cæca or tubuli; but when these bodies are aggregated into dense and solid masses with an extended system of excretory ducts, and when the whole of this apparatus is inclosed in a proper membrane so as to form a distinct body, such a body is termed a gland.

734. Primary aggregations of these secreting bodies constitute what is termed a conglobate,

that is, a simple gland; such are all the glands connected with the absorbent or lymphatic system. Secondary aggregates, or aggregates composed of simple glands, constitute what is termed a conglomerate, that is, a compound gland; such are all the organs commonly termed viscera, as the liver, the spleen, the pancreas, the kidney, and so on.

735. The conglobate, or simple gland, being formed by the aggregation of cryptæ, follicles, cæca, or tubuli, inclosed in a proper membrane, presents the appearance of a simple solid body, commonly of a rounded or oblong form (fig. CLXXVI. 516). On the contrary, the conglomerate or compound gland, being formed by the aggregation of conglobate or simple glands, presents the appearance of a compound body composed of a congeries of masses (fig. CLXV. 1). The larger masses enveloped in their own proper membrane are termed lobes (fig. CXCI.); the smaller masses, also enveloped in their own proper membrane, are termed lobules (fig. CXCI.); the lobules, when carefully examined, are seen to be composed of still smaller masses, and these of masses yet more minute, until at length patient, laborious, and skilful dissection brings into view the ultimate constituent elements, which are invariably found to consist of simple cryptæ, follicles, cæca, or tubuli.

736. Thus membrane having a specific arrangement of blood-vessels and nerves, from being simply extended, is folded into a few elementary forms; the bodies which result constitute simple secreting

organs; these bodies collected together form, by their aggregation, compound organs; the compound organs, uniting, form aggregates still more compound, until at length a structure is built up highly elaborate and complex. But this complexity of combination and arrangement does not alter the constitution of the organs; their form varies, but their nature remains essentially the same. All consist alike of membrane organized in a similar mode. The complex contains no element not possessed by the simple gland, and the gland contains no element not possessed by the secreting surface. But there is this difference in the complex organs. Every kind and degree of change in the form of the secreting apparatus, from membrane simply extended, to membrane coiled up into the most complex gland, is attended with an accumulation and concentration of secreting surface. The crypt contains a larger extent of secreting surface than the simple membrane; the follicle than the crypt; the cæcum than the follicle; and the tubulum than the cæcum. A certain amount of secreting surface is gained by the disposition of the simple membrane into the form of the crypt. The collection of a number of crypts into a cluster doubles the extent of the secreting surface by the extent of every crypt that is added to the cluster. The addition of every cluster doubles the whole extent of surface acquired by a single cluster. But when stems spring as if from a common trunk; when branches spring from a stem; when small branches spring from the large branches, and yet smaller branches from the small in a series, which the eye, assisted by the most powerful microscope, is wholly unable to trace; when all the clusters thus formed are collected, and combined into a compact mass, the intricacy of which no art can completely unravel, the extent of surface obtained is altogether immeasurable. How immense must be the extent of surface thus acquired in such an organ as the human lungs, in such a gland as the human liver!

737. In such an aggregation the concentration is also equal to the accumulation; the maximum of surface is comprised in the minimum of space, and the energy and elaborateness of the function of a secreting organ is uniformly proportionate to such a concentration of its secreting substance.

738. Hence the complexity of the compound gland in the higher animals would appear to arise solely from the intricate arrangement of the immense mass of secreting matter concentrated in a small compass; hence also the progressively increased complication indicated in the successive development of the glandular system in the animal series. Thus, for example, among the distinct organs formed for the purpose of elaborating a specific secretion, being intimately connected with the process of digestion, one of the first is the salivary gland. Low down in the scale, in the animal in which the first rudiment of a salivary gland is traceable, it consists of a single follicle, which

appears to serve the office of a gland. In an animal a little higher in structure, two, three, or four follicles combine to form a somewhat less simple organ. In an animal still higher in the series, a number of follicles are clustered together and form a much more complex organ; and in this manner, as the organization of the animal becomes higher and higher, the complexity of the gland increases, until at length it is composed of a countless number of follicles collected into clusters, the clusters disposed into lobes, the lobes subdivided into lobules, and the lobules into still smaller particles, the ultimate elements of the glandular apparatus. In like manner, when the first rudiment of the liver is discoverable, it consists of a single pouch or cæcum; somewhat higher in the series, the organ is composed of two or more cæca distinct and free; and then, as its complexity increases with the perfection of the organization, cæca are accumulated upon cæca; the aggregates so formed are closely compacted, disposed into lobes, divided into lobules, and subdivided into the ultimate particles of the glandular apparatus. So in a gland composed of tubuli, as the kidney, the organ in its rudimentary state consists of a few straight tubuli: as its structure advances more tubuli are added: next, the increasing tubuli superimposed one upon another become tortuous; then the tubuli still accumulating, become not merely tortuous, but convoluted; and last of all, countless numbers of tubuli are closely compacted into exceedingly convoluted masses. Uniformly, the lower the animal and the simpler the organ, the larger and the more manifest are the elementary parts of the gland; but in the higher animals these elementary bodies are so minute as to be altogether microscopical

Fig. CLXXXVIII.

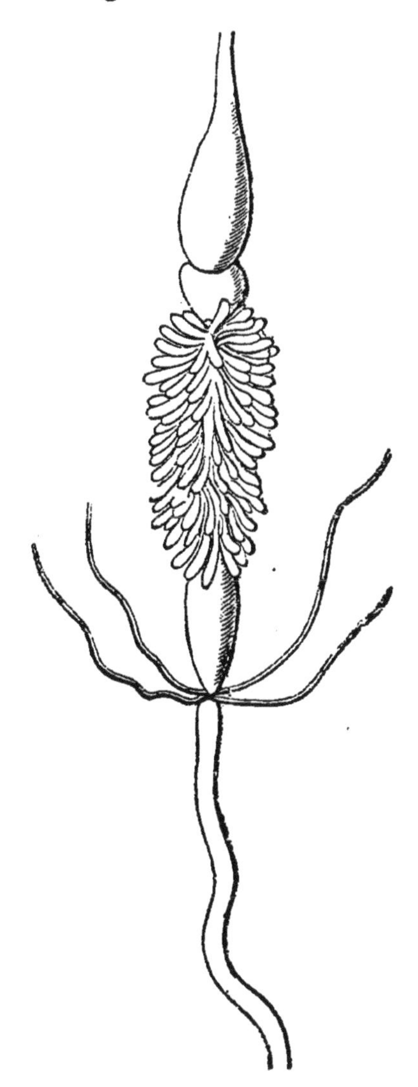

Aggregated and clustered cæca opening into the alimentary canal, performing the function of the liver.

Fig. CLXXXIX.

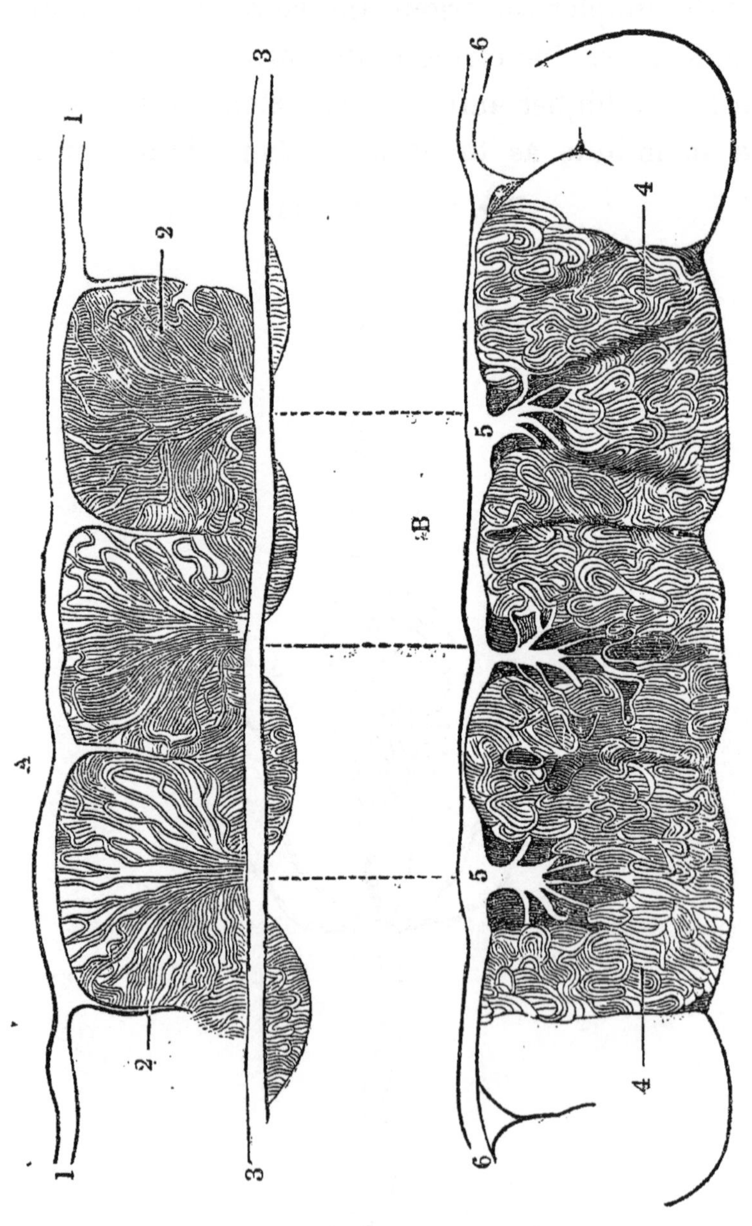

Portions of the kidney taken from the ophidian reptile, as seen under the microscope, highly magnified. A one portion of the kidney, showing—1. The trunk of the artery passing to be distributed to—2. The diverging tubuli, forming the uriniferous ducts which terminate in—3. The common excretory duct called ureter.—B another portion of the same kidney, showing the extremely convoluted course of—4. The uriniferous ducts. 5. The smaller excretory ducts, or secreting canals, converging and uniting to form—6. The common excretory duct called the ureter.

and their arrangement is so complex that it can be unravelled only with extreme difficulty.

739. It is a striking confirmation of the correctness of this view of the structure of the glandular apparatus, that whenever in the ascending series a gland appears for the first time in any class, the elementary bodies are so large, and are disposed in so simple a mode, that a slight examination is sufficient to demonstrate their primitive form, and to render it manifest that they consist either of vesicles, follicles, cæca, or tubuli, more or less aggregated. This is seen in the obvious structure presented by the liver, the pancreas, the salivary glands, and the mammæ, in the simple animals in which these organs first appear. Thus the liver in animals low down in the scale is manifestly composed of simple clustering follicles: in the fish the pancreas is composed of simple branched follicles: in the bird, the salivary glands are composed of simple parallel tubuli; and in the cetacea the breasts are composed of simple branched tubuli.

740. But the microscope, by bringing the successive development of the compound gland in the embryo of the higher animal under the cognizance of sense, perfectly discloses the nature of its composi-

tion. In the development of the incubated egg every step of the progressive formation of the compound gland is rendered visible to the eye. When this process is carefully watched, it is seen that the part of the gland first formed is the excretory duct, which springs from the blastema, the common mass of matter out of which all the organs are formed.

Fig. CXC.

A lobule of a gland in the progress of development in the ovum of the bird, as seen under the microscope, showing the origin of the excretory ducts in the semipellucid gelatinous blastema, and the branching and foliated arrangement of the follicles in which the excretory ducts terminate.

From this duct the elementary parts of the gland bud just as bunches of grapes bud from the stalk. The buds, at first at considerable distances from each other, approach nearer as they increase by new growths, until at length they come into actual contact. The growth continuing, and the compactness of the substance of the gland proportionally increasing, the primitive form of the elementary

Fig. CXCI.

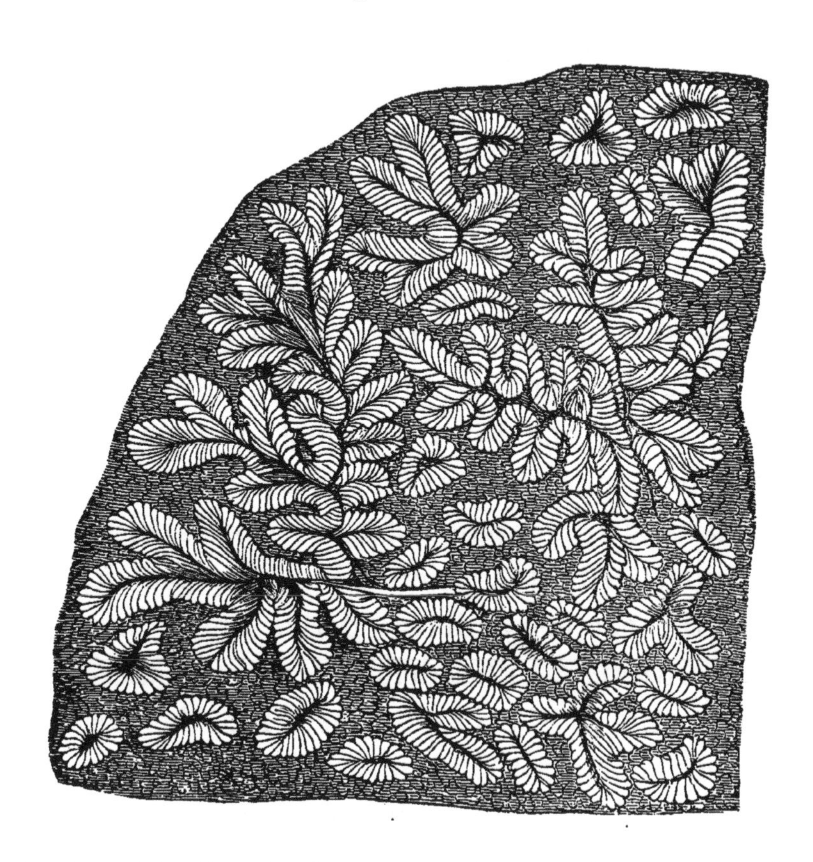

Section of the liver in the lower animal in the progress of development, as seen under the microscope, showing the rudimentary division into lobes and lobules, and the elongated terminations of the biliferous ducts, or cylindrical acini variously disposed in a branching and foliated manner.

bodies which compose it is ultimately lost. The substance of the gland now appears to consist of compact solid matter, which is commonly termed parenchyma. The component particles of this parenchymatous and apparently solid substance present a clustered or grape-like appearance, from which they early obtained the name of acini, from the Latin word acinus, a berry. This term, originally employed merely to express the clustered and branching appearance of the elementary parts of the gland, has since been used in widely different senses. By some it has been employed to express solid glandular grains constituting a supposed distinct parenchymatous substance, differing in every different gland. It is now proved that no such solid granular particles enter into the composition of any gland in the animal kingdom. By others the term acini has been employed to express granular bodies composed of blood-vessels, directly continuous with the excretory ducts, and from which the excretory ducts derive their origin. Recent investigation has demonstrated that there is no continuity of the blood-vessels into the excretory duct either in the acini or in any other part of the gland. It is established that the blood-vessels are spread out upon the walls of the secreting canals and do not form with them continuous tubes. The bodies which have been mistaken for granular particles, constituting the so called solid acini, are really the shut extremities of hollow follicles, cæca, or tubuli, which appear solid only from the closeness with which they are compacted. When carefully dissected and examined under the microscope, their real nature becomes apparent, and this is also sometimes capable of being demonstrated by injection; for some of these elementary bodies are vesicular, and can be filled with mercury, when they present a beautiful appearance like clusters of diamonds; or they may be inflated with air, just as the air vesicles of the lungs.

741. On watching the formation of the gland in the development of the embryo, it would appear that at first free streams of blood, or blood not contained in proper vessels, pass around the acini, the shut extremities of the excretory ducts, or the secreting canals. "So it would seem," says Müller, "when we examine the evolution of the liver and kidney in the embryo of the lower animal; for the interstices of the canals appear bloody, without the slightest trace of the walls of blood-vessels. I conceive that in the beginning new streams arise in an amorphous mass (a mass without form), not bounded by proper parieties; but that soon walls are formed, which present definite boundaries to the streams, the density of the substance around the streams gradually increasing." It is in this manner that the connexion is first established between the system of capillary blood-vessels and that of the secreting organs.

742. In its embryo state the compound gland of the highest animal consists of mere excretory ducts,

wonderfully similar to the simple secreting bodies of the lowest classes. But in the higher animal this simple form of the gland is transient: gradually, with the progressive evolution of the embryo, it passes into a more complex structure; while in the lower animal the simple form of the gland remains permanently the same through the whole term of life.

743. Such are the main points which have been ascertained relative to the structure of the secreting apparatus, which enters in one or other of its forms, as a constituent element, into almost every part of the animal body. Wherever there is nutrition there is secretion, and wherever there is secretion there is one or other of these secreting bodies. How immense the number of these organs in the human body! Every point in the interior of the walls that bound the great cavities is a secreting surface. Every point of the secreting surface that lines the alimentary canal, from its commencement to its termination, is studded with distinct secreting organs. Every point of the skin is still more thickly studded with distinct secreting organs. By the naked eye, and still more distinctly with a lens, may be seen the pores through which the vapour that constitutes the insensible perspiration incessantly exudes. Next are the open mouths of myriads of sebacious follicles that pour out upon the skin the oily matter which gives it its suppleness and softness;

and besides all these, are the hairs, each the product of a secreting organ placed immediately beneath the skin. An attempt to count the number of pores and hairs visible to the eye within the compass of an inch, and thence to compute the number on the whole surface of the skin, may convey some conception of the amount of these organs; yet these form but a small part of the secreting apparatus. The great viscera of the body, the brain, the lungs, the liver, the pancreas, the spleen, are portions of it; all the organs of the senses, the eyes, the ears, the nose, the tongue; all the organs of locomotion; every point of the surface of every muscle, and a great part of the surface and substance of the very bones are crowded with secreting organs.

744. Since every secreting organ is copiously supplied with blood, it follows that a great part of the blood of the body is always circulating in secreting organs; and, indeed, it is to afford materials for the action of these organs that the blood itself is formed.

745. How do these organs act upon the blood?

All that is known of the course of that portion of the blood which flows through an organ of secretion is, that it passes into arteries of extreme minuteness, which are spread out upon the external walls of the elementary secreting bodies, and which, as far as they can be traced, pass into capillary veins,—nowhere terminating by open mouths—nowhere presenting visible outlets or pores; their contents probably transuding through their thin and tender coats by the process of endosmose.

746. As it is flowing through these capillary arteries, the blood undergoes the transformations effected by secretion; forming—1. The fluids, which are added to the aliment, and which accomplish its solution, and change it into chyme. 2. The fluids, which are added to the chyme to convert it into chyle, and both to chyle and lymph, to assist in their assimilation. 3. The fluids which, poured into the cavities; facilitate automatic or voluntary movements. 4. The fluids, which serve as the media to the organs of the senses by which external objects are conveyed to the sentient extremities of the nerves for their excitement. 5. The fluids which, deposited at different points of the cellular tissue, when more aliment is received than is needed, serve as reservoirs of nutriment to be absorbed when more aliment is required than can be afforded by the digestive organs. 6. The fluids which are subsequently to be converted into solids. 7. The fluids which are eliminated from the common mass, whether of fluids or solids, to be carried out of the system as excrementitious substances. 8. In addition to all these substances, which are indispensable to the preservation of the individual, those which are necessary to the perpetuation of the species.

747. In order to form any conception of the mode in which the secreting organs act upon the blood, so as to elaborate from it such diversified substances, it is necessary to consider the chemical composition of the different products of secretion, and the degrees in which they really differ from each other, and form the common mass of blood out of which they are eliminated.

748. By chemical analysis, it is established that all the substances which are formed from the blood by the process of secretion are either water, albumen, mucus, jelly, fibrin, oil, resin, or salts ; and, consequently, that all the secretions are either aqueous, albuminous, mucous, gelatinous, fibrinous, resinous, oleaginous, or saline.

749. 1. AQUEOUS SECRETIONS.—From the entire surface of the skin, and also from that of the lungs, there is constantly poured a quantity of water, derived from the blood, mixed with some animal matters, which, however, are so minute in quantity, that they do not communicate to the aqueous fluid any specific character.

750. 2. ALBUMINOUS SECRETIONS.—All the close cavities, as the thorax, the abdomen, the pericardium, the ventricles of the brain, and even the interstices of the cellular tissue, are constantly moistened by a fluid which is termed serous, because it is derived from the serum of the blood. This serous fluid consists of albumen in a fluid form, and it differs from the serum of the blood

chiefly in containing in equal volumes a smaller proportion of albumen. Membranes of all kinds consist essentially of coagulated albumen; and the albumen, as constituting these tissues, differs from albumen as existing in the serum of the blood only in being unmixed with extraneous matter, and in being in a solid form.

751. 3. Mucous Secretions.—As all the close cavities, or those which are protected from the external air, are moistened with a serous fluid, so all the surfaces which are exposed to the external air, as the mouth, the nostrils, the air-passages, and the whole extent of the alimentary canal, are moistened with a mucous fluid. Mucus does not exist already formed in the blood. It is always the product of a gland. Some of the mucous glands are among the most elaborate of the body; still the main action of the gland seems to be to coagulate the albumen of the blood, for the basis of mucous is coagulated albumen. The fluid that lubricates the mucous surfaces in their whole extent, the saliva, the gastric juice, the tears, the essential part of the fluid formed in the testes and in the ovaria, are mucous secretions. Hence the most complex and elaborate functions of the body, respiration, digestion, reproduction, are intimately connected with the mucous secretions: nevertheless, as far as regards their chemical nature, the mucous differ but slightly from the albuminous secretions; and it is probable that a slight change

in the secreting organ is sufficient to convert the one into the other. By the irritation of mercury on the salivary glands, the saliva, properly of a mucous, is sometimes converted into a substance of an albuminous nature; and irritation in some of the serous membranes occasionally causes them to secrete a mucous fluid.

752. 4. GELATINOUS SECRETIONS.—The proximate principle termed jelly abounds plentifully in several of the solids of the body, and more especially in the skin; but jelly does not exist already formed in the blood. Yet it is not the product of a gland, neither is there any known organ by which it is formed. Out of the body albumen is capable of being converted into jelly by digestion in dilute nitric acid: this conversion is probably effected by the addition of a portion of oxygen to the albumen. Albumen contains more carbon and less oxygen than jelly; the proportions of hydrogen and nitrogen in both being nearly the same. According to MM. Gay Lussac and Thénard, the elements of albumen and jelly are,

	Carbon.	Oxygen.	Hydrogen.	Nitrogen.
Albumen ...	52.883	23.872	7.54	15.765
Jelly	47.881	27.207	7.914	16.988

The conversion of albumen into jelly is incessantly going on in the system; and the process accomplishes most extended and important uses. In the lungs at the moment of inspiration oxygen enters into the blood in a state of loose combination; but in the system, at every point where the conversion of albumen into jelly takes place, oxygen probably enters into a state of chemical combination with albumen; and the new proximate principle, jelly, is the result. The agent by which this conversion is effected appears to be the capillary artery: the primary object of the action is the production of a material necessary for the formation of the tissues of which jelly constitutes the basis, as the skin; but a secondary and most important object is the production of animal heat; the carbon that furnishes one material of the fire being given off by the albumen at the moment of its transition into jelly; and the oxygen that furnishes the other material of the fire being afforded to the blood at the moment of inspiration. This view affords a beautiful exposition of the reason why jelly forms so large a constituent of the skin in all animals. The great combustion of oxygen and carbon, the main fire that supports the temperature of the body, is placed where it is most needed, at the external surface.

753. 5. Fibrinous Secretions. The pure muscular fibre, or the basis of the flesh, is identical with the fibrin of the blood. It contains a larger proportion of nitrogen, the peculiar animal principle, and is consequently more highly animalized than the preceding substances. It appears to be simply discharged from the circulating blood by the

capillary arteries, and deposited in its appropriate situation; no material change in its constitution being, it would seem, necessary to fit it for its office.

754. 6. Oleagenous Secretions.—Fat of all kinds, which is found so extensively connected with the muscles, and with many of the viscera, and which is more or less diffused through the whole extent of the cellular tissue, marrow, milk, and nervous and cerebral matter, are essentially of the same nature. the basis of them all is oil; and oil exists already formed both in the chyle and in the blood.

755. 7. Resinous Secretions.—The peculiar substance forming the basis of bile, picromel; the peculiar substance forming the basis of urine, urea; the peculiar substance connected with the muscular fibre, and forming a component part of almost all the solids and fluids of the body, osmazome, consists of a common principle—a resin, which exists already formed in the blood, and more especially in the serosity of the blood.

756. 8. Saline Secretions.—The substances termed saline, namely, the acids, the alkalis, and the neutral and earthy salts, are disposed over every part of the system: they enter more or less into all the constituents both of the solids and fluids; they form more especially the phosphate of lime, the earthy matter of which bones are composed; and they all exist already formed in the blood.

757. From this account, then, it appears, that by chemical analysis, the blood is ascertained to contain water, albumen, fibrin, oil, resin, and various saline and earthy substances: it follows, that, with the exception of the absence of jelly, the constituents of the body and the constituents of the blood are nearly identical; and it is probable that they will be found to be perfectly identical when their analysis shall have become complete.

758. It is also manifest that in by far the greater number of cases the various substances of which the body is composed are simply separated from the nutritive fluid at the parts of the body at which they are deposited; and that, existing already formed in the blood, they are merely deposited there, and not generated. Still, however, since it is certain that gelatin cannot be recognized in the blood, and since it is doubtful whether some other substances found in different textures and secretions really exist in the blood, it is necessary, in the present state of our knowledge, to suppose, that although most of the constituents of the living tissues are contained in the blood, yet that in some instances a material change is effected in their nature at the time and place of their escape from the circulation; and that in these cases the secreted substances are not simple extracts from, but products of, the blood.

759. It is by the apparatus of secretion that this

separation, evolution, or re-formation, is effected Out of a fluid which contains, blended together, almost all the heterogeneous substances of which the body is built up, particular substances are selected from the common mass, and are deposited in certain parts, and only in certain parts. Although by the most careful examination of the structure of the apparatus, it is not possible to form a precise conception of the mode in which this separation is effected, yet we are enabled to perceive a number of contrivances which we can readily understand must conduce to the accomplishment of the object.

760. 1. Of these, the most obvious is mechanical arrangement.

761. In its passage to different organs the blood is propelled through canals of extreme minuteness: in every different case these canals differ from each other in size; pass off from their respective trunks at different angles; possess different degrees of density; are variously contorted, and are of various lengths. In some they are straight, in others convoluted; at one time branching, at another pencillated, and at another starry. The veins, too, in some cases, are almost straight, in others exceedingly tortuous, in others reticulated; and the freedom of their communication with the arteries varies so much, that in some cases fine injections pass from the one set of vessels to the other with the greatest facility, while, in others

they pass with extreme difficulty. The consequence of these divers arrangements of the capillary blood-vessels is, that the current of the blood must necessarily flow in them with different degrees of velocity; its particles must be placed at different distances from each other, and must be presented to each other in different positions and in widely different proportions. In no two secreting organs are any two of these conditions exactly alike. In the lower orders of animals, in which secretion is seen in its simplest condition, the general nutritive fluid, elaborated and contained in a single internal cavity, appears to furnish a variety of products very different from itself, by a process hardly more complex than mere transudation through a living membrane. In the higher animals the different secreting organs may be considered, in part at least, as mechanical contrivances adapted to carry on analogous transudations—fine sieves or strainers diversly constructed. A fluid containing such heterogeneous matters as the blood, held in combination by so slight an affinity, slowly transuding through series of tubes, the mechanical arrangement of which is so varied, must yield a different substance in every different case. Thus by simply filtering the blood a vast variety of products may be obtained, merely in consequence of a varied disposition of the minute tubes of which the filters are composed.

762. 2. But in the second place, this diversity

of mechanical arrangement is calculated in a high degree to promote and to modify chemical action. The contact or proximity of the particles of bodies, the extent of surface which those particles present to each other, the space of time in which they continue in contact, the degree of force with which they impinge against each other, the degree of temperature to which they are exposed, —these, and circumstances such as these, are conditions which exert the most powerful influence over chemical decomposition and re-combination. In the different secreting organs, as has been shown, the blood must necessarily pass through vessels having every conceivable diversity of diameter: in those vessels it must consequently flow with corresponding differences of velocity. Some of these diameters will admit one constituent of the blood, as one of the red particles; others may be large enough to admit two or more of the red particles abreast; others may be so small as to be incapable of admitting a single red particle, receiving only the more fluid portions of the blood; in some vessels these different constituents will be in one degree of proximity, in others in another; in some they will remain long in contact, in others only for an instant: it is obvious that from such different conditions the chemical products may be infinitely varied.

763. Such is the composition of chemical bodies, that a great diversity of substances is obtainable

merely by changing one condition, the proportions in which the elementary particles combine.

764. Oxygen and nitrogen combined in one proportion form atmospheric air; in another proportion, nitrous oxide; in another, nitric oxide; in a fourth, nitrous acid; and in a fifth, nitric acid. Few secretions formed from the blood differ more widely from each other than the products thus formed from these two elementary bodies.

765. Urea consists of two prime equivalents of hydrogen, one of carbon, one of oxygen, and one of nitrogen. Remove one of the atoms of hydrogen, and take away the atom of nitrogen, urea is converted into sugar; combine with urea an additional atom of carbon, it is changed into lithic acid. In like manner add a small quantity of water to farina, it is converted into sugar; to fibrin, it is changed into adipocere. From a reservoir containing a quantity of substances in the state of vinous fermentation, draw off portions of the liquor at different stages of the process, and cause these to pass through tubes of various diameters and with various degrees of velocity, there will be obtained at one time an unfermented syrup, at another, a fermenting fluid, at another, wine, at another, vinegar. Out of the body place the blood in a state of rest, it will spontaneously separate into serum and crassamentum, and the crassamentum will further separate into fibrin and red particles. Add to the serum a certain portion of acid,

it will be coagulated into solid albumen; add to this solid albumen another portion of acid, it will be converted into jelly. Add a certain portion of acid to fibrin, it will be changed into adipose matter; bring the acid into contact with the red particles, they will be converted into a substance closely resembling bile. If by the rough chemistry which the art of man can conduct so great a variety of substances may be obtained out of a single compound, it is not wonderful that a far greater variety should be produced by the delicate and subtle chemistry of life.

766. 3. But a third most important agent in the process of secretion is some influence derived from the nervous system.

1. It is proved, by direct experiment, that the destruction of the nervous apparatus, or of any considerable portion of it, stops the process of secretion. By experiments performed by Mr. Brodie, it is ascertained that the secretion of the urine is suspended by the removal or destruction of the brain, though the circulation be maintained in its full vigour by artificial respiration.

2. The section, and still more the removal, of a portion of the sentient nerves of the stomach (the par vagum, or eighth pair), according to some experimentalists, deranges and impedes; according to others, totally arrests the process of digéstion.

3. Other classes of phenomena illustrate in a striking manner the influence of the nervous

system over the process of secretion. The sight, nay, even the thought of agreeable food, increases the secretions of the mouth. Pleasurable ideas excite, painful ideas destroy, the appetite for food; probably, in the one case, by increasing, and, in the other, by suspending the secretion of the gastric juice: the emotion of grief instantly causes a flow of tears; that of fear, of urine; the sight or thought of her child fills the maternal breasts with milk, while the removal of the child from the mother diminishes and ultimately stops the secretion.

767. Even the imagination is capable of exerting a powerful influence over the process. A female who had a great aversion to calomel was taking that medicine in very small doses for some disease under which she was labouring. Some one told her that she was taking mercury: immediately she began to complain of soreness in the mouth; salivated profusely, and even put on the expression of countenance peculiar to a salivating person. On being persuaded that she had been misinformed, the discharge instantly began to diminish, and ceased altogether in a single night. Two days afterwards she was again told, on good authority, that calomel was contained in her medicines, upon which the salivation immediately began again, and was profuse. That this salivation was not produced by the calomel, but was the effect solely of the influence of imagination on the salivary glands,

was proved by the absence of redness of the gums, which always takes place in mercurial salivation, and also by the absence of the peculiar fætor, which is characteristic of the action of this metal on the system.

768. The same influence is apparent even in the lower animals: exhibit food to a hungry dog, the saliva will pour from its mouth. Rob the nest of the bird of its eggs as soon as they are laid, the bird may be made to deposit eggs almost without end, though if the eggs are allowed to remain undisturbed, it will lay only a certain number. The bird is led by instinct to continue to deposit eggs in the nest until a certain number is accumulated; that is, a mental operation acts upon the ovarium, the secreting organ in which the eggs are formed, maintaining it in a state of active secretion for an indefinite period; whereas without that mental operation the secretion would be limited to a definite number.

769. In all these cases it is probable that the vital agent by which the effect is produced on the secreting organs is the organic nerve. Though the sentient part of the nervous system may in many cases be the part primarily acted on, yet there is reason to believe that the ultimate effect is invariably produced on the organic part, the sentient nerves in this case acting on the organic, as in other cases the organic act on the sentient, in consequence of that intimate connexion which, for

the reason assigned (vol. i. p. 79), is established between both parts of this system. For,

770. 1. The true object of the sentient part of the nervous system is to establish a relation between the body and the external world; the object of the organic part is to preside over the functions by which the body is sustained and nourished, that is, over the processes of secretion.

771. 2. The nerves which are distributed to the secreting arteries, and which increase in number and size as the arteries become capillary, are, for the most part, derived from the organic portion of the nervous system (fig. CLXX. 3). This anatomical arrangement clearly points to some physiological purpose, and indicates the closeness of the relation between the function of the organic nerve and the ultimate action of the capillary artery.

772. 3. It is demonstrated that the sentient part of the nervous system, though occasionally influencing and modifying secretion, is not indispensable to it. In tracing the normal or regular development of the human fœtus, it is found that the heart is constructed and is in full action before the brain and spinal cord, the central masses of the sentient part of the nervous system, are in existence; and that these masses are themselves built up by processes to which the action of the heart is indispensable; consequently, innumerable acts of secretion must have taken place, those, for

example, which have been necessary to form the different substances which enter into the composition of the heart, before the brain and spinal cord exist. In like manner in the anormal or irregular development of the fœtus, as in the production of monsters, there may be not a vestige of head, neck, brain or spinal cord, while there may be a perfect heart, perfect lungs, perfect intestines, and various portions even of the osseous system.

773. However in the perfect animal secretion may be under the influence of the brain and spinal cord, it is clear that, since the process can go on without them, it must be independent of them. It is a false induction from these facts drawn by some physiologists that secretion is independent of the nervous system. They do prove that it is independent of one part of the nervous system, the sentient; but it does not follow that it is independent of the other part, the organic.

774. 4. It is demonstrated that the organic part of the nervous system is not only independent of the sentient part, but that it is even pre-existent to it. Researches into the development of the nervous system, as shown in the progressive growth of the fœtus of different animals, have proved that the existence of the organic nerves is manifest long before that of the sentient; that nerves are discoverable in the tissues, before the brain and the spinal cord are formed; that as these masses become visible and grow, nerves springing from the

tissues advance towards the central nervous masses, and at length unite with them; but that this union does not take place until the development of the nervous system is considerably advanced. These curious and most instructive facts show that in the fœtus, though the brain and spinal cord may have been destroyed or have been non-existent, yet that the organic nerves may have been in full action. After a communication has been once established between the two parts of the system, indeed, the destruction of the brain or spinal cord may stop secretion, not because these organs are indispensable to secretion; but because the destruction of one part of the system involves the death of the other, just as the organic life itself perishes soon after the destruction of the animal.

775. The existence of the organic nerve is probably simultaneous with that of the secreting artery: from the first to the last moment of life the nerve regulates the artery; the influence of the one is indispensable to the operation of the other; and, by their conjoint action, the sentient nerve itself, as well as every other organ, is constructed.

776. There is reason to believe that the physical agent by which the organic nerve influences secretion is electricity. The nerve appears to be the medium by which electrical fluid is conveyed to the secreting organs, and the nerve probably influences secretion by influencing chemical combination, through the intervention of this most

powerful chemical agent. This is rendered probable by the observation of various phenomena, and by the result of direct experiment.

777. 1. It is proved that galvanic phenomena may be excited by the contact of the nerve and muscle in an animal recently dead. A galvanic pile may be constructed of alternate layers of nervous and muscular substance, or of nervous substance and other animal tissues. A secreting organ liberally supplied with organic nerve is probably then in its physical structure nothing but a galvanic apparatus. It is certain that some animals, as the raia torpedo, possess a special electrical apparatus composed essentially of nervous matter; that the nerves which compose this apparatus correspond strictly with the organic nerves of the human body; that they are distributed principally to the organs of digestion and secretion, and that they exert a powerful influence over these processes; for, when the animal is frequently excited to give shocks, digestion appears to be completely arrested; so that, after the animal's death, food swallowed some time previously is found wholly unchanged.

778. 2. It is universally admitted that the nerves in all animals possess an extreme sensibility to the stimulus of electricity, and more especially to that form of it which is termed galvanism.

779. 3. Direct experiment proves that the stimulus of galvanism may be made to produce in the

living body precisely the same effect as the nervous influence. It has been stated, that the division of the par vagum, in the neck of a living animal, suspends the digestion of the food probably by stopping indirectly the secretion of the gastric juice. If after the division of the nerves, their lower ends, that is, that portion of the nerves which is still in communication with the stomach, but no longer in communication with the brain, be made to conduct galvanic fluid to the stomach, secretion goes on as fast as when the nerves are entire and conduct nervous influence. Dr. Wilson Philip having divided the par vagum in the neck of a living animal, coated a portion of the lower end of the nerves with tin foil, placed a silver plate over the stomach of the animal, and connected respectively the tin and silver with the opposite extremities of a galvanic apparatus. The result was that the animal remained entirely free from the distressing symptoms which had always before attended the division of the nerves, and that the process of digestion, which had been invariably suspended by this operation, now went on just as in the natural state of the stomach. On examining the stomach after death, the food was found perfectly digested, and afforded a striking contrast to the state of the food contained in the stomach of a similar animal, in whom the nerves had been divided, but which had not been subjected to the galvanic influence.

780. 4. On applying a low galvanic power to a saline solution contained in an organic membrane, Dr. Wollaston found that the galvanic fluid decomposed the saline solution, and that the component parts of the solution transuded through the membrane; each constituent being separately attracted to the corresponding wire of the interrupted circuit. This experiment, says this acute and philosophical physiologist, illustrates in a very striking manner the agency of galvanism on the animal fluids. Thus the quality of the secreted fluid may probably enable us to judge of the electrical state of the organ which produces it; as for example, the general redundance of acid in urine, though secreted from blood that is known to be alkaline, appears to indicate in the kidney a state of positive electricity; and since the proportion of alkali in bile seems to be greater than is contained in the blood of the same animal, it is not improbable that the secretory vessels in the liver may be comparatively negative.

781. We may imagine, says Dr. Young, that at the division of a minute artery a nervous filament pierces it on one side, and affords a pole positively electrical, and another opposite filament a negative pole. Then the particles of oxygen and nitrogen contained in the blood, being most attracted by the positive point, tend towards the branch which is nearest to it; while those of the hydrogen and carbon take the opposite channel; and that both

these portions may be again subdivided, if it be required; and the fluid thus analysed may be recombined into new forms by the reunion of a certain number of each of the kinds of minute ramifications. In some cases the apparatus may be somewhat more simple than this; in others, perhaps, much more complicated; but we cannot expect to trace the processes of Nature through every particular step; we can only inquire into the general direction of the path she follows.

782. Considerations such as these afford us a glimpse into the mode in which Nature conducts some of her most secret and subtile operations; or rather into the immediate agency by which she effects them; for, properly speaking, of the mode in which she works, we do not obtain the slightest insight, and even of her immediate agency our view, at least in the present state of our knowledge, is indistinct and vague. By the study of the apparatus which she builds up, we can trace back her operations a step or two; but in every case, at a certain point, the apparatus itself becomes so delicate as to elude our senses, and then of course we are necessarily at a stand. So, the rough materials with which she carries on her great work of secretion, by careful analysis we can separate into divers parts, and ascertain that each part possesses peculiar properties. The main channels by which she conveys these varied constituents to the different parts of the system

we can trace; the delicate organs by which she produces on these rude materials her wonderful transformations we can see; but beyond the threshold of these organs we cannot go. Why from one common mass of fluid the same variety of peculiar substances are constantly separated, and each in its respective place: why the kidney never secretes milk, nor the liver urine, nor the breast bile: why membrane, and muscle, and bone, and fat, and brain, are uniformly deposited in the same precise situation: why these depositions go on with uniformity, constancy and regularity; and by what laws each process is controlled and modified, we do not know. But though with whatever diligence we investigate these operations, the great problem remains, and probably ever will remain unresolved, still it is both a pleasurable and a profitable labour to follow Nature in her path, to the extreme point to which it is possible to trace her footstep; for the phenomena themselves are often in the highest degree curious and interesting; while their order and relation can seldom be so considered as to be understood, without the suggestion of practical applications of great and permanent usefulness.

CHAPTER XII.

OF THE FUNCTION OF ABSORPTION.

Evidence of the process in the plant, in the animal—Apparatus general and special—Experiments which prove the absorbing power of blood-vessels and membrane—Decomposing and analysing properties of membrane—Endosmose and exosmose—Absorbing surfaces, pulmonary, digestive, and cutaneous—Lacteal and lymphatic vessels—Absorbent glands—Motion of the fluid in the special absorbent vessels—Discovery of the lacteals and lymphatics—Specific office performed by the several parts of the apparatus of absorption—Condition of the system on which the activity of the process depends—Uses of the function.

783. Absorption is the function by which external substances are received into the body, and the component particles of the body are taken up from one part of the system, and deposited in some other part. So universal and constant is the operation, that there is not a fluid nor a solid, not a surface nor a tissue, not an external nor an internal organ, which is not, in its turn, the seat and the subject of the process. By its action the component particles of the living body are kept in a state of perpetual mutation.

784. The plant in a humid atmosphere increases in weight. The nutritive matter of the plant diffused in the soil is taken up by its capillary rootlets, or by the spongolæ which are attached to them, and conveyed into the system. The fall of dew or rain upon leaves promotes the growth of the plant. Leaves placed on water are capable of preserving not only their own vitality, but that of the branches and twigs to which they are attached. These phenomena show that the process of absorption is carried on by the plant.

785. The evidence of the absorbing power possessed by the animal is still more striking.

786. 1. If an animal be immersed in water the amount of which is ascertained by measure, its head being kept out of the water, so that none can enter the mouth, the body increases in weight and the water diminishes in quantity. If certain animals, as snails, are plunged in water impregnated with colouring matter, the fluids in the interior of their body soon acquire the colour of the water by which they are surrounded. Frogs, previously kept for some time in dry air, when placed in water, absorb a quantity equal in weight to their whole body.

787. 2. In a humid atmosphere the animal increases in weight still more than the plant.

788. 3. If a quantity of water be injected into any of the great cavities of the body, as into that of

the peritoneum, the whole of the fluid after a certain time disappears; it is spontaneously removed.

789. 4. If in the progress of disease a fluid be poured into any cavity of the body, as often happens in dropsy, the whole of the fluid is removed, sometimes spontaneously and quite suddenly; but more often slowly, under the influence of medicinal agents.

790. 5. Certain substances, whether applied to an external or an internal surface, produce specific effects on the system, just as when they are received into the stomach or injected into the blood-vessels. Mercury in mere contact with the skin, but more rapidly when the application is aided by friction, produces the same specific action upon the salivary glands, and the same general action upon the system as when the preparation of the metal is received into the stomach. By the like external and local application arsenic, opium, tobacco, and other narcotics produce their distinct and peculiar effects on the nervous system, and their remote and general effects on the other systems.

791. 6. If an organ or tissue be deprived of nourishment, it gradually diminishes in bulk, and at length wholly disappears from the system. By long-continued pressure, such as that occasioned by the pulsation of a diseased artery, as in aneurism, or by the growth of a fleshy tumor, portions of the

firmest and strongest muscle, nay, even of the most dense and compact bone, wholly disappear. At one time the fluids diminish in quantity, the flesh wastes, and the weight of the body is reduced one half or more. Under other circumstances, while the state of the general system remains stationary, some particular part diminishes in size, or altogether disappears.

792. 7. Healthy and strong men, engaged in hard labour and exposed to intense heat, sometimes lose, in the space of a single hour, upwards of five pounds of their weight. Though daily engaged for months together in this occupation at two different periods of the day, for the space of an hour each time, and though consequently these men lose five pounds twice every day, yet when weighed at intervals of three, six, or nine months, it is found that the weight of the body remains stationary, not varying, perhaps, more than a pound or two. It follows that the bodies of these men must absorb, twice every day, a quantity equal in weight to that which they lose.

793. These phenomena depend on a power inherent in the body, that of taking up and carrying into the system certain substances in contact with its surfaces, and of transporting from one part of its system to another its own component particles.

794. The apparatus by which these operations are carried on is general and special.

795. The general apparatus consists of blood-

vessels and membrane. The special apparatus consists of a peculiar system of vessels, namely, the lacteals and lymphatics, together with the system of glands termed conglobate.

796. It is proved by direct experiment that the walls of blood-vessels exert a power by which substances in contact with their external surface penetrate their tissue, reach their internal surface, and mix with the mass of the circulating fluids, and that this property is possessed by all blood-vessels, arteries and veins, great and small, dead and living.

797. If a portion of a vein or artery taken from the body be attached by either extremity to two glass tubes in order to establish a current of warm water in its interior, if the vein be then placed in a fluid slightly acidulated, and the fluid which flows through the vessel be collected in a flask, this latter fluid becomes, in the space of a few minutes, sensibly acid. In this experiment there is no possibility of communication between the current of warm water and the external acidulated fluid, consequently the latter must penetrate the parietes of the vessel, that is, absorption must take place through its membranous walls.

798. A striking experiment demonstrates the absorbing power of the living blood-vessels. If the trunk of a vein or artery be exposed in a living animal, and a poisonous substance in solution be dropped on the external surface of either, the animal is

killed in a few minutes, just as when the poison is injected into the blood-vessel itself. Analogous experiments on the minute blood-vessels not only show that they are endowed with the like absorbing power, but that their number, tenuity and extent, are conditions which greatly favour the activity of the process.

799. Membrane is an organised substance abounding with blood-vessels. Whether the absorbing power possessed by this tissue be due only to these vessels, or whether it be assisted in the operation by other agents not yet fully ascertained, it is certain that the absorbing power it exerts is highly curious and wonderful.

800. An animal membrane placed in contact with water becomes saturated with fluid : placed in contact with a compound fluid, as with water or spirit holding colouring matter in solution, the membrane actually decomposes the compound and resolves it into its elementary parts, just as accurately as can be done by the chemist. If one extremity of a piece of membrane be placed in a vessel containing the tincture of iodine, for example, and the other extremity be kept out of the fluid, that portion of the membrane which is in immediate contact with the tincture acquires a perfectly dark colour, because the iodine completely penetrates the substance of the membrane. This dark-coloured portion is bounded by a definite line, above which the membrane is penetrated by

a different part of the solution, by a pearly colourless fluid, the alcohol in which the iodine was suspended. Above this again there are traces of a still lighter coloured fluid, which is probably water. In like manner, if strips of membrane are placed in glasses containing port wine, the same analytical process is effected by the membrane. The colouring matter of the wine is imbibed by the lower portion of the membrane; above this is the alcohol, and above this the water.

801. These and many analogous experiments demonstrate that the process of absorption is accompanied with the further phenomena of decomposition and analysis; and that membrane, at the very moment it imbibes certain compound substances, resolves them into their constituent elements.

802. It is further established by numerous experiments that different compound substances are decomposed and absorbed by membrane with different degrees of facility. If strips of membrane are placed in phials containing different kinds of fluids, one fluid rises only a line or two; others rise to the height of many inches. There is indubitable evidence that analogous properties are possessed by living membrane; that the mucous membrane of the stomach at the moment it imbibes, decomposes and analyses the alimentary and medicinal substances in contact with its surface; and consequently that in all animals mem-

brane becomes a most important agent in carrying on the digestive process.

803. But perhaps the most remarkable property possessed by membrane is that of establishing in fluids in contact with its surfaces currents through its parietes, which proceed in opposite directions, according to the different natures of the fluids, and more especially according to their different densities. If small bladders composed of membrane are filled with a fluid of greater density than water, and securely fastened, and then thrown into water, they acquire weight and become swollen and tense. If the experiment be reversed; if the bladders be filled with water and immersed in a denser fluid, the denser fluid flows inwards to the water, and the water passes from the interior outwards. M. Dutrochet, who was led by accident to the observation of these phenomena, and who saw at once the possible importance of this agency in some organic processes hitherto involved in great obscurity, commenced an extended series of experiments with a view to ascertain the exact facts. He took the cæca of fowls, membranous bags already made to his hand, into which he introduced a quantity of fluid consisting of milk, thin syrup, or gum-arabic dissolved in water. Having securely tied the membranes, he placed the bags thus filled in water, and found that two opposite currents are established through the walls of the cæca. The first and strongest current,

that from without inwards, is formed by the flow of the external water towards the thicker fluid contained in the cæca; the second and weaker current, that from within outwards, is formed by the flow of the thicker interior fluid towards the external water. The first or the in-going current is termed *endosmose*, from ενδον, intus, and ωσμος, impulsus, and the second or out-going current is termed *exosmose*, from a similar combination of Greek words signifying an impulse outwards.

804. The velocity and strength of these currents are capable of exact admeasurement. The amount of endosmose is measured by an apparatus termed an endosmometer, which consists of a small bottle, the bottom of which is taken out and the aperture closed by a piece of bladder. Into this bottle is poured some dense fluid; the neck of the bottle is closed with a cork, through which a glass tube, fixed upon a graduated scale, is passed. The bottle is then placed in pure water. The water by endosmose penetrates the bottle in various quantities according to the density of the fluid contained in its interior through the membrane closing its bottom. The dense fluid in the bottle, increased in quantity by the addition of the water, rises in the tube fitted to its neck, and the velocity of its ascent is the measure of the velocity of the endosmose.

805. The strength of endosmose is measured by a similar apparatus, in which a tube is twice

bent upon itself, and the ascending branch containing a column of mercury which is raised by the fluid in the interior of the endosmometer, as the volume of this fluid is increased by the endosmose. By means of these two instruments it is found that the velocity and strength of endosmose follow the same law, and that both are proportionate to the excess of the density of the fluid contained in the endosmometer above the density of water. By numerous experiments it is ascertained that by employing syrup of ordinary density (I. 33) an endosmose is obtained, the strength of which is capable of raising water more than 150 feet.

806. But though difference of density is necessary to the production of endosmose, yet numerous and decisive experiments show that the different natures of fluids, irrespective of their proportionate densities, materially influence the activity and energy of the process. Thus, if sugar-water and gum-water of the same density be placed in the same endosmometer, the former produces endosmose with a velocity as seventeen and the latter only as eight. The endosmose produced by a solution of the sulphate of soda is double that produced by a solution of the hydro-chlorate of soda of the same density. A solution of albumen exerts an endosmose four times greater than a solution of gelatin of the same density.

807. With organic fluids endosmose goes on without ceasing until the chemical nature of the

fluids becomes altered by putrefaction; but with alkalies, soluble salts, acids, and chemical agents in general, the endosmose excited is capable only of short continuance, because such agents enter into chemical combination with the organic tissue of the endosmometer, and thus destroy endosmose.

808. It is remarkable that the direction of the endosmotic currents produced by vegetable membrane is the exact reverse of that produced by animal membrane under precisely the same circumstances. Thus oxalic acid, when separated from water by an animal membrane, invariably exhibits endosmose from the acid towards the water; when separated by a vegetable membrane, from the water towards the acid: and the same is the case with the tartaric and citric acids, and with the sulphuric, the hydro-sulphuric, and the sulphurous acids. I filled, says Dutrochet, a pod of the *colutea arborescens*, which being opened at one end only, and forming a little bag, was readily attached by means of a ligature to a glass tube, with a solution of oxalic acid, and having plunged it into rain-water, endosmose was manifested by the ascent of the contained acid fluid in the tube, that is to say, the current flowed from the water towards the acid. The lower part of the leek (*allium porrum*) is enveloped or sheathed by the tubular petioles of the leaves. By slitting these cylindrical tubes down one side, vegetable membranous webs of sufficient breadth and strength to be tied upon the

reservoir of an endosmometer are readily obtained. An endosmometer, fitted with one of these vegetable membranes, having been filled with a solution of oxalic acid and then plunged into rain-water, the included fluid rose gradually in the tube of the endosmometer, so that the endosmose was from the water towards the acid, the reverse of that which takes place when the endosmometer is furnished with an animal membrane. Vegetable membrane, then, at least with fluids containing a preponderance of acid, produces a current, the direction of which is the exact reverse of that produced by animal membrane.

809. The bodies of organised beings are composed in great part of various fluids of different density, saparated from each other by thin septa, precisely the conditions which are necessary to the production of endosmose. But such conditions never concur in inorganic bodies, whence inorganic bodies never exhibit endosmotic phenomena. Vegetable tissue of every kind consists of vast multitudes of aggregated cells intermingled with tubes. The parietes of these hollow organs are exceedingly delicate and thin; the organs themselves are at all times filled with fluids, the densities of which are infinitely various; consequently, by endosmose and exosmose, mutual interchanges of their contents incessantly go on; those contents brought into contact by currents moving now in one direction and now in another now rapidly and now slowly intermingle, and in

consequence of their admixture changes in their chemical composition take place. It is by these powers that water holding in solution nutrient matter diffused through the soil penetrates the spongeolæ of the capillary rootlets, always filled with a denser fluid than the water contained in the soil,—that the energetic motion by which the sap ascends is generated,—that the ascending sap is attracted into fruits, always of greater density than the crude sap,—that buds are capable of emptying the tissue that surrounds them when they begin to grow, and that almost all the phenomena connected with the motions of fluids in plants, and the chemical changes which those fluids undergo in consequence of this admixture, is effected. And there cannot be a question that analogous phenomena take place in the various cells, cavities, and minute capillary vessels of the animal body.

810. It is then established on indubitable evidence that all animal tissues, without exception, possess an inherent property by which they are capable of transmitting through their substance certain fluids, and even solids, convertible into fluids; and that the great agent by which this transmission is effected is membranous tissue, whether in the form of blood-vessels or of proper membrane. By virtue of this property fluids and solids are absorbed, by the animal body, with whatever surface or organ they are in contact, whether with an external or an internal surface, or with the eye, the mouth,

the tongue, the stomach, the lungs, the liver, or the heart.

811. But membrane is so disposed and modified, in different parts of the body, as to admit of the introduction of fluids and solids from the exterior to the interior of the system with widely different degrees of facility. There may be said to be in the human body three great absorbing surfaces, the pulmonary, the digestive, and the cutaneous, each highly important, but each endowed with exceedingly different degrees of absorbing power.

812. The pulmonary surface, for reasons which will be readily understood from what has been already stated relative to the structure of the air vesicles of the lungs, is by far the most active absorbing surface of the body. The mode in which the air vesicles are formed and disposed has been shown to be such as to give to the lungs an almost incredible extent of membranous surface, while the membrane of which the cells are composed is exceedingly fine and delicate. Moreover, there is the freest possible communication between all the branches of the pulmonary vascular system, whether arteries or veins; the distance between the lungs and the heart is short; the course of the blood from the pulmonary capillaries to the central engine that works the circulation is rapid, and the lungs are at the same time close to the central masses of the nervous system, with which indeed they are placed in

direct communication by nerves of great magnitude and of most extensive distribution. These circumstances account for the wonderful rapidity with which substances are absorbed, when placed in contact with the pulmonary surface, and for the instantaneousness and intensity of the impression produced upon the system, when the substance thus introduced is of a deleterious nature.

813. They also afford an explanation of a phenomenon not to have been credited without experience of the fact, that innoxious substances introduced into the air cells of the lungs in moderate quantities produce no more inconvenience there than when taken into the stomach. A single drop of pure water, when in contact near the glottis with the same membrane that forms the air vesicles of the lungs, excites the most violent and spasmodic cough, and the smallest particle of a solid substance permanently remaining there occasions so much irritation that inevitable suffocation and death result. Yet so different is the sensibility of this membrane in different parts of its course, that while at the upper portion of the trachea it will not bear a drop of water without exciting violent disturbance, in the air vesicles it tolerates with only slight inconvenience a considerable quantity even of solid matter. An accident of a nature sufficiently alarming, which occurred to Dessault, affords a striking illustration of this curious fact. This celebrated surgeon had

to treat a case in which the trachea and esophagus were cut through. It was necessary to introduce a tube through the divided esophagus into the stomach, and to sustain the patient by food introduced in this manner. On one occasion the tube, instead of being passed through the esophagus to the stomach, was introduced into the trachea down to the division of the bronchi. Several injections of soup were actually thrown into the lungs before the mistake was discovered; yet no fatal, and even no dangerous consequences ensued. Since that period, in various experiments on animals, several substances of an innoxious nature have been thrown into the lungs without producing any inconvenience beyond slight disturbance of the respiration and cough. The reason is, that after a short time the substances are absorbed by the membrane composing the air vesicles, and are thus removed from the lungs and borne into the general circulating mass. At every point of the pulmonary tissue there is a vascular tube ready to receive any substance imbibed by it, and to carry it at once into the general current of the circulation.

814. Hence the instantaneousness and the dreadful energy with which poisons and other noxious substances act upon the system when brought into contact with the pulmonary tissue. A solution of nux vomica injected into the trachea produces death in a few seconds. A single inspi-

ration of the concentrated prussic acid kills with the rapidity of a stroke of lightning. This acid in its concentrated form is so potent a poison, that it requires the most extreme care in the use of it, and more than one physiologist has been poisoned by it through the want of proper precaution while employing it for the purpose of experiment. If the nose of an animal be slowly passed over a bottle containing this poison, and the animal happen to inspire during the moment of the passage, it drops down dead instantaneously, just as when the poison is applied in the form of liquid to the tongue or the stomach. The vapour of chlorine possesses the property of arresting the poisonous effects of prussic acid, unless the latter be introduced into the system in a dose sufficiently strong to kill instantly; and, hence, when an animal is all but dead from the effects of prussic acid, it is sometimes suddenly restored to life by holding its mouth over the vapour of chlorine.

815. Examples of the transmission of gaseous bodies through the pulmonary membrane have been already fully described in the account of the passage of atmospheric air to the lungs, and of carbonic acid gas from the lungs, in natural respiration. But foreign substances may be mixed with or suspended in the atmospheric air, which it is the proper office of the pulmonary membrane to transmit to the lungs, and may be immediately carried with it into the circulating mass. Thus,

merely passing through a recently-painted chamber gives to the urine the odour of turpentine. The vapour of turpentine diffused through the chamber is transmitted to the lungs with the inspired air, and passing into the circulation through the pulmonary membrane, exhibits its effects in the system more rapidly than if it had been taken into the stomach, and thence absorbed.

816. Vegetable and animal matter in a state of decomposition generates a poison, which when diffused in the atmosphere, and transmitted to the lungs in the inspired air, produces various diseases of the most destructive kind. The exhalations arising from marshes, bogs, and other uncultivated and undrained places, constitute a poison of a vegetable nature, which produces principally intermittent fever or ague. Exhalations accumulating in close, ill-ventilated, and crowded apartments in the confined situations of densely-populated cities, where no attention is paid to the removal of putrefying and excrementitious matters, constitute a poison chiefly of an animal nature, which produces continued fever of the typhoid character. It is proved by fatal experience that there are situations in which these putrefying matters, aided by heat and other peculiarities of climate, generate a poison so intense and deadly that a single inspiration of the air in which they are diffused is capable of producing instantaneous death; and that there are other situations in

which a less highly concentrated poison accumulates, the inspiration of which for a few minutes produces a fever capable of destroying life in from two to twelve hours. In dirty and neglected ships, in which especially the bilge-water is allowed to remain uncleansed; in damp, crowded, and filthy gaols; in the crowded wards of ill-ventilated hospitals filled with persons labouring under malignant surgical diseases, or some forms of typhus fever, an atmosphere is generated which cannot be breathed long, even by the most healthy and robust, without producing highly dangerous fever.

817. The true nature of these poisonous exhalations is demonstrated by direct experiment. If a quantity of the air in which they are diffused be collected, the vapour may be condensed by cold and other agents, and a residuum of vegetable or animal matter obtained, which is found to be highly putrescent, constituting a deadly poison. A minute quantity of this concentrated poison applied to an animal previously in sound health, destroys life with the most intense symptoms of malignant fever. If, for example, ten or twelve drops of a fluid containing this highly putrid matter be injected into the jugular vein of a dog, the animal is seized with acute fever; the action of the heart is inordinately excited, the respiration is accelerated, the heat increased, the prostration of strength extreme, the muscular power

so exhausted, that the animal lies on the ground wholly unable to stir or to make the slightest effort; and, after a short time, it is actually seized with the black vomit, identical, in the nature of the matter evacuated with that which is thrown up by an individual labouring under yellow fever. It is possible, by varying the intensity and the dose of the poison thus obtained, to produce fever of almost any type, endowed with almost any degree of mortal power. These facts, of which practical applications of the highest utility are hereafter to be made, may suffice to show the importance of the pulmonary membrane as an absorbing surface. By the extent and energy of its absorbing power, it is one of the great portals of life and health, or of disease and death.

818. The digestive surface is of much less extent than the pulmonary; it is less vascular; it is further removed from the centre of the circulating system, and it is covered with a thick mucus, which is closely adherent to it; hence its absorbing power is neither so great as that of the pulmonary membrane, nor do noxious substances in contact with it affect the system so rapidly. An appreciable interval commonly elapses between the introduction of a poison into the stomach and its action upon the system. An emetic is commonly a quarter of an hour before it begins to operate: arsenic itself is generally half an hour, and sometimes three quarters of an hour, be-

fore it produces any decided effect on the system: but at length a noxious substance, applied to any part of the digestive membrane is introduced into the circulating mass and produces its appropriate effects on the system, just as when it is in contact with the pulmonary tissue.

819. Over the external surface of the body or the skin, there is spread a thin layer of solid, inorganic, insensible matter, like a varnish of Indian rubber. The obvious effect of such a barrier placed between the external surface of the body and external objects, is to moderate the entrance of substances from without, and the transmission of substances from within, that is, to regulate both the absorbing and the exhaling power of the skin. Hence the comparative slowness with which substances enter the system by the cutaneous surface; the impunity with which the most deadly poisons may remain for a time in contact with the skin, with which prussic acid, arsenic, corrosive sublimate, may be touched and even handled. The internal surface of the body is protected from the action of acrid substances introduced into the alimentary canal by a layer of mucus through which an irritant must penetrate before it can pain the sentient nerve or irritate the capillary vessel; but were not a still denser shield thrown over the external surface, pain, disease, and death must inevitably result from the mere contact of innumerable

bodies, which now are not only perfectly innoxious, but capable of ministering in a high degree to human comfort and improvement.

820. Immediately beneath the cuticle is a surface as vascular as it is sensitive, from which absorption takes place with extreme rapidity. Poison in very minute quantity introduced beneath the cuticle kills in a few minutes. Arsenic applied to surfaces from which the cuticle has been removed by ulceration produces its poisonous effects upon the system just as surely as when introduced into the stomach. The poisonous matter of small-pox and of cow-pox placed in almost inappreciable quantity by the lancet beneath the cuticle produces in a given time its specific action upon the system. When, in certain states of disease, with the view of bringing the system rapidly under the influence of a medicinal agent, the cuticle is removed by a blister, and the exposed surface is moistened with a solution of the substance whose action is required, the constitutional effects are developed with such intensity, that if extreme care be not taken in the employment of any deleterious substance in this mode the result is fatal in a few minutes.

821. The phenomena which have been stated may suffice to illustrate the absorbing power of the general tissues and surfaces of the body; but superadded to this, there is carried on in particular

parts of the system a specific absorption for which a special apparatus is provided.

822. The special apparatus of absorption, commonly termed the proper absorbent system, consists of the lacteal and lymphatic vessels and of the conglobate glands. The lacteals arise only from the intestines; the lymphatics, it is presumed, from every organ, tissue, and surface of the body. Both sets of vessels possess a structure strikingly analogous to that of veins, the common agents of absorption. The coats of the lacteals and lymphatics are somewhat thinner and a good deal more transparent than those of veins; yet thin and delicate as they are, they possess considerable strength, for they are capable of bearing, without rupture, injections which distend them far beyond their natural magnitude.

Fig. CXCII.

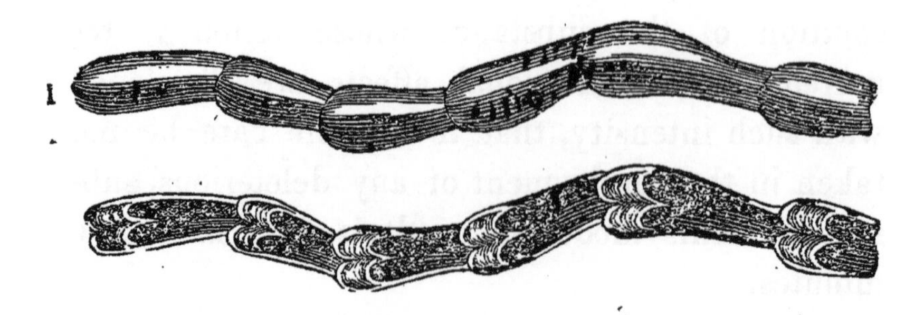

An enlarged view of an absorbent vessel.—1. External surface, with the jointed appearance produced by the valves.—2. The same vessel laid open, showing the arrangement of the valves.

823. When fully distended, these vessels present a jointed appearance somewhat resembling a string of beads (fig. CXCII. 1). Each joint indicates the situation of a pair of valves (fig. CXCII. 2). These valves are of a semilunar form, and are composed of a fold of the inner coat of the vessel (fig. CXCII. 2). The convex side of the valve, in the lacteals, is towards the intestines; in the lymphatics towards the surfaces; in both towards the origins of the vessels. The valves allow the contents of the vessels to pass freely towards the main trunk of the system, but prevent any retrograde motion towards the origins of the vessels.

824. By continued pressure the resistance of the valves may be overcome, so that mercury may be made to pass from the trunk into the branches. When this is done in an absorbent trunk proceeding from certain organs, such as the liver, it is seen that the absorbents are distributed, arborescently, in such vast numbers that the surface of the viscus appears as if it were covered with a reticular sheet of quicksilver.

825. The internal coat of the small intestines has been shown to present a fleecy surface, crowded with minute elevations called villi, which give this surface an appearance closely resembling the pile of velvet. Each villus consists of an artery, a vein, a nerve, and a lacteal, united and sustained by delicate cellular tissue. After a meal the lac-

teals become so turgid with chyle that they completely conceal the blood-vessels and nerves, so that the surface of the intestine presents to the eye only a white mass, or a surface thickly crowded with white spots (fig. CXCIII.)

Fig. CXCIII.

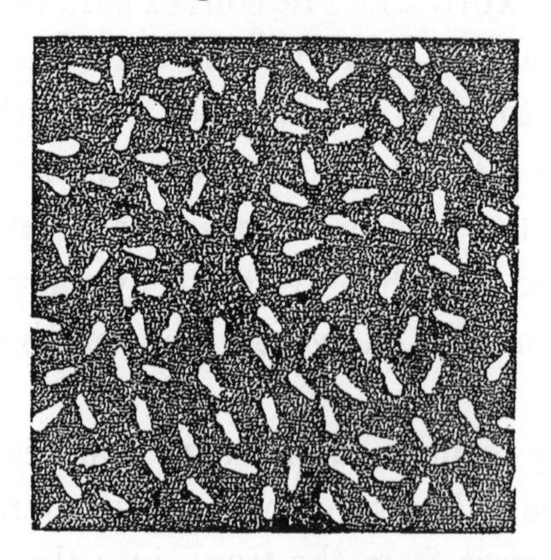

Appearance of the lacteals turgid with chyle, as seen in the jejunum some time after a meal.

826. When a portion of the intestine in this condition of the lacteal vessels is examined under the microscope, there is said to be visible on the

Fig. CXCIV.

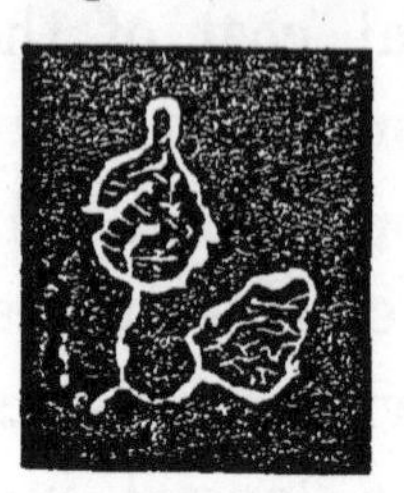

Magnified view of two ampullulæ turgid with chyle, terminating the lacteal vessels.

villus an oval vesicle, termed an ampullula (fig. CXCIV.). This vesicle is described as having an aperture at its apex, which it is conceived constitutes the open mouth of the lacteal, and through which the chyle is supposed to be taken up.

827. Mr. Cruikshank, who particularly devoted himself to the study of this part of the absorbent system, states that he had an opportunity of examining these vessels in a person who died suddenly some hours after having taken a hearty meal, and who had been previously in sound health. "In some hundred villi," he says, "I saw the trunk of the lacteal beginning by radiated branches (fig. CXCV.). The orifices of these radii were very distinct on the surface of the villus as well as the

Fig. CXCV.

View of villi, with the lacteals arising from their surface by open mouths and forming radiated branches. The surface of one of these villi is represented as entirely white, from the lacteals being so turgid with chyle as completely to obscure their orifices and their radiating branches.

radii themselves (fig. cxcv.). There was but one trunk in each villus. The orifices on the villi of the jejunum, as Dr. Hunter said (when I asked him as he viewed them in the microscope how many he thought there might be) were about fifteen or twenty in each villus, and in some I saw them still more numerous" (fig. cxcv.).

828. The course of the lacteals, from their origin in the villi to their termination in the thoracic duct, has been traced (687). It is conjectured that the lymphatics take their origin from every point of the body, but it is admitted that they have not been actually seen even in every organ; still they have been found in so many that it is inferred that they really exist in all, and that in those in which they have not been hitherto detected they have eluded observation on account of their extreme delicacy and transparency and our imperfect means of examining them.

829. Though, like veins, lymphatics anastomose freely with each other, yet they do not proceed from smaller to larger branches and from larger branches to trunks, but continue of nearly the same magnitude from their origin to their termination. They are disposed in two sets, one of which always keeps near the external surface of the body, and the other is deeply seated, accompanying more especially the great trunks of the blood-vessels.

830. In the human body every vessel that can

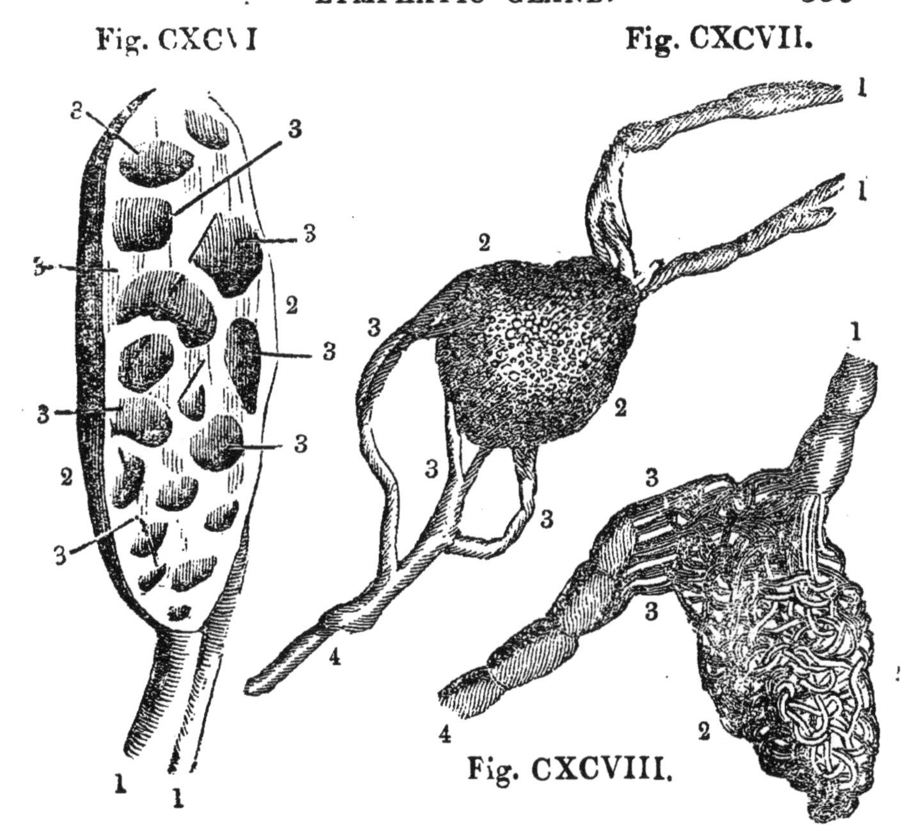

CXCVI.—1. Trunks of absorbent vessels entering a gland. 2. Gland laid open. 3. Highly magnified views of the cells or follicles of which the gland is supposed to consist. CXCVII. — 1. Absorbent vessels called vasa inferentia, entering (2) the gland. 3. Absorbent vessels emerging from the gland, called vasa efferentia, and forming (4) a common trunk. CXCVIII.—1. Trunk of absorbent vessel entering a gland. 2. Gland apparently composed entirely of convoluted vessels. 3. Vessels emerging from the gland and forming (4) a common trunk.

be distinctly recognised either as a lacteal or a lymphatic, passes, in some part of its course, through a conglobate or lymphatic gland (figs. CXCVII., CXCVIII.). These glands, small, flattened, circular or oval bodies, resembling beans in shape, are enclosed in a distinct membranous envelope. Their intimate structure has been already fully described (chap. xi.). They are of various sizes, ranging from three to ten lines in diameter: they are placed in determinate parts of the body, and are grouped together in various ways, being sometimes single, but more often collected in masses of considerable magnitude. Numerous absorbent vessels, termed vasa inferentia, enter the gland on the side remote from the heart (figs. CXCVII. 1 and CXCVIII. 1); a smaller number, called vasa efferentia, leave it on the side proximate to the heart (fig. CXCVII. 3). If mercury be injected into the vasa inferentia (fig. CXCVI.), it is seen to pass into a series of cells of the corresponding gland (fig. CXCVI. 3), and then to escape by the vasa efferentia; but if the gland be more minutely injected, as by wax, all appearance of cells vanishes; the whole substance of the gland seems then to consist of convoluted absorbents (fig. CXCVIII. 2), irregularly dilated, and communicating with each other so intimately that every branch that leaves the gland appears to have been put in communication with every branch that entered it (fig. CXCVIII. 1, 2, 3).

831. The motion of the fluid within the absorbent vessels, though not rapid, is energetic. If a ligature be placed around the thoracic duct in a living animal, the tube will swell and ultimately burst, from the rupture of its coat, in consequence of the force of the distension that

takes place below the ligature. If the thoracic duct in the neck of a dog be opened some hours after the animal has taken a full meal, the chyle flows from the vessel in a full stream, and in the space of five minutes half an ounce of the fluid may be obtained. Yet this system of vessels is beyond the influence of the circulating blood: it has no heart to propel it; no current behind always in rapid motion to urge it onwards; it is therefore inferred that it is moved by a vital contractile power inherent in the vessels, analogous to, if not identical with, muscular contractility. The flow of blood through the arterial tubes is universally believed to be effected, in part at least by such a contractile power, for this, among other reasons, that if in a living animal the trunk of an artery be laid bare, the mere exposure of it to the atmospheric air causes it to contract to such a degree that its size becomes obviously and strikingly diminished (298.1). The same phenomenon has been observed in the main trunk of the absorbent system. Tiedemann and Gmelin state that in the course of their experiments they saw the thoracic duct contract from exposure to the air

832. The delicacy and transparency of the lacteals and lymphatics long concealed them from the view of the anatomist. The lacteals had indeed been occasionally seen in ancient times, but their office was altogether unknown. In the

year 1563 Eustachius discovered the thoracic duct, but did not perceive its use. About half a century afterwards, in the year 1622, the lacteals were again one day by chance seen by Asellius, in Italy, while investigating the function of certain nerves. Mistaking the lacteals for nerves, he at first paid no attention to them; but soon observing that they did not pursue the same course as the nerves, and "astonished at the novelty of the thing," he hesitated for some time in silence. Resolving in his mind the doubts and controversies of anatomists, of which it chanced that he had been reading the very day before, in order to examine the matter further, "I took," he says, "a sharp scalpel to cut one of these chords, but scarcely had I struck it when I found a liquor white as milk, or rather like cream, to leap out. At this sight I could not contain myself for joy; but turning to the by-standers, Alexander Tadinus and the senator Septalius, I cried out Εὕρηκα! with Archimedes; and at the same time invited them to look at so rare and pleasing a spectacle, with the novelty of which they were much moved. But I was not long permitted to enjoy it, for the dog now expired, and, wonderful to tell, at the same instant the whole of that astonishing series and congeries of vessels, losing its brilliant whiteness, that fluid being gone, in our very hands, and almost before our eyes, so evanished and disappeared that hardly a vestige was left to my most diligent search." The next day he procured another dog, but could not discover the smallest white vessel. "And now," he continues, "I began to be downcast in my mind, thinking to myself that what had been observed in the first dog must be ranked among those rare things which, according to Galen, are sometimes seen in anatomy." But at length recollecting that the dog had been opened "athirst and unfed," he opened a third "after feeding him to satiety; and now everything was more manifest and brilliant than in the first case." The zeal with which he followed out the clue he had obtained is indicated by the number of dogs, cats, lambs, hogs, and cows which he dissected, and by the statement that he even bought a horse and opened it alive; but, he adds, "a living man, how ever, which Erasistratus and Herophilus of old did not fear to anatomize, I *confess* I did not open."

833. Nearly thirty years elapsed before the lacteals, which were long thought to terminate in the liver, were traced to the thoracic duct; and it was not until the year 1651, about eighty years after the discovery of Asellius, that the lymphatics were discovered, and that the whole of this portion of the absorbent system was brought to light.

834. Taking together the whole of the appa-

ratus of absorption, the specific office performed by its several parts seems to be as follows:—

835. 1. It is established that the lacteals absorb chyle, and that they refuse to take up almost every other substance which can be presented to them. Experimentalists are uniform in stating that however various the substances introduced into the stomach, it is exceedingly rare to find in the lacteals anything but chyle. These vessels appear to be endowed with a peculiar sensibility, derived from the nervous system, by which they are rendered capable of exerting an elective power, readily absorbing some substances and absolutely rejecting others.

836. 2. The lymphatics absorb a far greater variety of substances than the lacteals, but not all substances indiscriminately; chiefly organized matter in a certain stage of purification; particles passing through successive processes of refinement (707).

837. 3. The blood-vessels, and more especially the capillary veins, appear to absorb indiscriminately all substances, however heterogeneous their nature, which are dissolved or dissolvable in the fluids presented to them.

838. 4. The absorbent glands appear by various modes, either by removing superfluous and noxious matters, or by the addition of secreted substances possessing assimilative properties, to

approximate the fluid which flows through them more and more closely to the nature of the blood. Fatal effects result from the artificial infusion of minute portions even of mild substances into the blood. Hence the extended and winding course which Nature causes the new matter formed from the food to undergo, even after its elaboration in the digestive apparatus, in order that, before it is allowed to mingle with the blood, its perfect purification and assimilation may be secured.

839. The activity or inactivity of the process of absorption is mainly dependant on the emptiness or the plethora of the system. There is a point of saturation beyond which the absorbent vessels, though in immediate and continued contact with absorbable matters, will take up no more The nearer the system to this point the less active the process; the further the system from this point the more active the process. Thus, when an animal whose vessels are full to saturation is immersed in water, or exposed to humid air, its body does not increase in weight, and there is no sensible diminution of the water; but the longer an animal is kept without fluid, and the more it is exposed to the action of a dry air, the further its system is removed from the point of saturation, and exactly in that proportion, when it is brought in contact with water, is the diminution of the quantity of the fluid and the increase in the weight of the body. This law explains many

circumstances of the animal economy,—why it is impossible to dilute the blood or any other animal fluid beyond a certain point, by any quantity of liquid which may be in contact with the external surface, or which may be taken into the stomach; why it is impossible to introduce nutrient matter into the system, beyond a certain point, by any quantity of food, which the digestive organs may convert into chyle; why, consequently, the bulk and weight of the body are incapable of indefinite increase; why that bulk and weight are so rapidly regained after long abstinence; and why the appetite is so keen, and the ordinary fulness and plumpness of the body are so soon restored, after recovery from fever and other acute diseases, when the digestive organs have been uninjured.

840. Different portions of the absorbent apparatus accomplish specific uses. With the absorbent action of the capillary blood-vessels and of membranous surfaces every organic function, but more especially the processes of digestion and respiration, are intimately connected.

841. The specific absorption carried on by the lacteals has for its object the introduction of new materials into the system, for the reparation of the losses which it is constantly sustaining by the unceasing actions of life.

842. The specific absorption carried on by the lymphatics has a twofold object. First, the introduction of particles, which have already formed an integrant part of the system, a second time into

the blood, in order to subject them anew to the process of respiration, thereby affording them a second purification, and giving them new and higher properties ; and, secondly, the regulation of the growth of the body, and the communication and preservation of its proper form.

843. It is the office of the lacteals to replenish the blood by constantly pouring into it new matter, duly prepared for its conversion into the nutritive fluid. It is the office of the lymphatics to preside over the distribution of the blood as it is deposited in the system in the act of nutrition. The lymphatics are the architects which mould and fashion the body. They not only regulate the extension of the frame, but they retain each individual part in its exact position, and give to it its exact size and shape. Growth is not mere accretion, not simple distension; it consists of a specific addition to every individual part, while all the parts retain the same exact relation to each other and to the whole. When a bone grows it does not increase in bulk by the mere accumulation of bony matter; but every osseous particle is so increased in length and breadth that the relative size of every part, and the general configuration of the whole organ, remain precisely the same. When a muscle grows, while the entire organ enlarges in bulk by the augmentation of every individual part, each part retains exactly its former proportions and

its relative connexions. When the brain grows a certain quantity of cerebral matter is added to every individual part, but at the same time the proportionate size and original form of each part, and the primitive configuration of the entire organ, are retained exactly the same. How is this effected? By a totally new disposition of every integrant particle of every part of every organ. New matter is not deposited before the removal of the old: the lymphatic, in the very act of removing the old, fashions a mould for the reception of the new, and then the capillary artery brings the new particle and deposits it with unerring exactness in the bed prepared for it. Thus, by removing the old materials of the body in a determinate manner, and thereby fashioning a mould for the reception of the new, the lymphatics may be said, in the strictest sense, to be the architects of the frame.

CHAPTER XIII.

OF THE FUNCTION OF EXCRETION.

In what excretion differs from secretion—Excretion in the plant—Quantity excreted by the plant compared with that excreted by the animal—Organs of excretion in the human body—Organization of the skin—Excretory processes performed by it—Excretory processes of the lungs—Analogous processes of the liver—Use of the deposition of fat—Function of the kidneys—Function of the large intestines—Compensating and vicarious actions—Reasons why excretory processes are necessary—Adjustments.

844. THE various matters contained in organized bodies, and even those which enter as constituent elements into their composition, are constantly removed from the system, and thrown off into the external world. The matters thus rejected are called excretions; and the various processes by which their elimination is effected constitute a common function termed excretion.

845. Excretion is the necessary consequence of the deterioration which all organized matter undergoes by the actions of life. The matters removed by the process consist of the waste particles of the body, or the particles expended in the

vital actions, as the aliment contains the particles which replenish the waste, and compensate the expenditure.

846. The excretions are separated from the common organized mass by processes perfectly analogous to those comprehended in the great function of secretion. Excretion is only a particular form of secretion: the difference between the two functions is, that, in the former, the matter eliminated being either noxious or useless, is separated for the sole purpose of being rejected; while, in the latter, the matter eliminated is destined to perform some useful purpose in the economy. Accordingly, the products of excretion are termed excrementitious; and those of secretion, recrementitious.

847. The chief matters excreted by the plant are oxygen, carbonic acid, air, water, in some few cases, under peculiar circumstances, ammonia and chlorine; and in still rarer cases, during the night, poisonous substances, as carburetted hydrogen, together with acrid, and even narcotic principles.

848. The forms under which these excretions are eliminated are exceedingly various. Sometimes the matter excreted is in the shape of gas, at other times it is in that of vapour, and at others in that of liquid. The chief gaseous exhalations are oxygen and carbonic acid; the vaporous exhalations consist principally of water, in the

state of vapour; and the liquid exhalations are either pure water, or water holding in combination sugar, mucilage, and other proximate vegetable principles. Even the peculiar products formed by the vital actions of the plant, as the volatile oils, the fixed oils, the balsams, the resins, and perhaps, with the exception of gum, sugar, starch, and lignine, all the substances formed out of the proper juices of the plant, are true excretions; for these substances are fixed immovably in the cells, sacs, or tubes which secrete and contain them: they are not consumed in the growth of the plant; they do not appear to be applied to any useful purpose in the economy; they are injurious, and even poisonous to the very plant in which they are formed when taken up by the roots and combined with the sap: as long as they remain in the plant they are isolated in the individual parts in which they are first deposited, until with the advancing age of the plant they lose their aqueous particles, and are finally dried up; they, therefore, possess all the essential characters of excrementitious substances.

849. The organs by which these matters are excreted are the leaves, the flowers, the fruits, the roots, and certain bodies called glands.

850. The gaseous and vaporous exhalations are effected chiefly by the leaves, which it has been shown (320 and 465), under the influence of the solar ray, are always pouring out a large

quantity of oxygen, and still larger quantities of fluid in the state of vapour.

851. Similar matters are exhaled by the flowers either in the form of vapour or of liquid; and this exhalation commonly bears with it a peculiar odour, which proceeds from an essential oil, sometimes evaporated with the pollen, and at other times secreted by glandular bodies which have their seat in the petals.

852. Fruits, and especially green fruits, as raspberries, pears, apples, plums, apricots, figs, cherries, gooseberries, and grapes, pour out oxygen during the day, and carbonic acid gas during the night, and thus co-operate with leaves in carrying on the function of excretion.

853. The more elaborate excretions contained in special receptacles, and formed by diverse organs from the proper juices of the plant, descend chiefly by the bark, and are poured by the roots into the soil. These excretions, if re-absorbed by the roots, and re-introduced into the system of the plant that has rejected them, poison that plant. Consequently, two processes of deterioration are always going on in the soil; first, the absorption of the nutrient matter contained in it; and, secondly, the accumulation of excrementitious matter constantly poured into it by the growing plant. By the addition of manure, the soil is replenished with fresh nutritive materials; by a rotation of crops, it is purified from noxious ex-

cretions. It is a remarkable and beautiful adjustment, that excrementitious substances which are destructive to plants of one natural family, actually promote the growth of plants of a different species. Thus, if wheat be sown upon a tract of land proper for that grain, it may produce a good crop the first, the second, and perhaps even the third year, as long as the ground is what the farmers call in good heart. But, after a time, it will yield no more of that particular kind of corn. Barley it may still bear, and, after this, oats, and perhaps after these, pease, or some other species belonging to a different family. The excrementitious matter deposited in the soil by a preceding is absorbed by a succeeding crop; the matter excreted by the former serving as nutriment or stimulus to the latter. But though in this mode all noxious matter is removed from the soil, yet the ground at last becomes quite barren, in consequence of having parted with all its nutrient particles, and then it will yield no more produce until it is supplied with a new fund of matter. This new matter is afforded by vegetable or animal substances, in which, the principle of life having become extinct, the peculiar bond that held their particles together is dissolved. Leaves, flowers, fruits, bark, roots; hair, skin, horns, hoofs, fat, muscle, bone, the blood itself, whatever has formed a part of the organized body, now dead, and repassing through the process of decomposition, back to the simple physical elements, all its forms of beauty gone, and exhaling only matters highly deleterious to animal life, mixed with the soil, are recombined into new products, spring up into new plants, and thus re-appear under new forms of beauty, and afford fresh nutriment to myriads of animals. The very refuse of the matters which have served as food and clothing to the inhabitants of the crowded city, and which, allowed to accumulate there, taint the air, and render it pestilential, promptly removed, and spread out on the surface of the surrounding country, give it healthfulness, clothe it with verdure, and endow it with inexhaustible fertility.

854. The quantity of matter excreted by the plant is proportionate to the energy of its vital actions. Hence it is always greatest in spring, when the tender leaves are beginning to shoot; gradually diminishes as autumn approaches; and, at last, as the leaves turn yellow, and the vessels which connect the leaves with the stalk dry up and are closed, it almost wholly ceases.

855. It is copious in proportion to the number of the leaves, and to the extent of the surface they present. From experiments performed as long ago as the year 1699, by Woodward, it appears that, of the whole quantity of water absorbed by the plant, the least proportion exhaled to that retained is as 46 or 50 to 1; in many cases it is as 100 or 200 to 1, and in some above 700 to 1.

In one experiment, a plant which imbibed 2501 grains of water, increased in weight only three grains and a half: hence the dampness and humidity of the air in all places in which trees and the larger vegetables abound; more especially when the leaves are young, and most numerous and active; and hence also the magnitude of the rivers in all extensive countries which are covered with forests.

856. Exhalation, scarcely appreciable in the night, is most abundant during the day under the influence of the solar light. If two plants of the same size are covered with two glass bells, and one be exposed to the sun's light, while the other is left in the shade, the inner surface of the former bell becomes covered with drops of water, while that of the second remains perfectly dry.

857. The absolute quantity of matter excreted by the plant is widely different in different species. According to Hales, in a sun-flower three feet and a half high, the leaves of which presented a surface of 5616 square inches, or 39 square feet, the greatest quantity exhaled in twelve hours, during the day, was one pound fourteen ounces avoirdupois; the medium quantity one pound four ounces. In a middle-sized cabbage, the greatest quantity exhaled was one pound nine ounces; the medium quantity one pound three ounces. In a vine, the greatest quantity exhaled was six ounces; the medium quantity five ounces.

In a young apple tree having 163 leaves, the surface of which was equal to 1589 square inches, or 11 square feet, the greatest quantity exhaled was eleven ounces; the medium quantity nine ounces. Martino calculated the quantity exhaled by a cabbage, in the twenty-four hours, at twenty-three ounces; by a young mulberry-tree, eighteen ounces; and, by a maize plant, seven drachms.

858. Supposing the weight of the human body to be 160 pounds, and the weight of a sun-flower 3 pounds, the relative weights of the two bodies will be as 160 to 3, or as 53 to 1. The surface of such a human body is equal to 15 square feet, or 2160 square inches; the surface of the sun-flower is 5616 square inches, or as 26 to 10. The quantity perspired in the twenty-four hours by an ordinary-sized man, according to the estimate of Keill, is about thirty-one ounces. Allowing two ounces for the exhalation during the beginning and the ending of the night, the quantity exhaled by the plant, in the same time, is twenty-two ounces; so that the perspiration of a man to that of a sun-flower is nearly as 141 to 100, though the weight of the man to that of the sun-flower is as 53 to 1. Taking bulk for bulk, the plant imbibes seventeen times more fresh fluid than the man, partly, no doubt, for the reason assigned by Hales—because, "the fluid which is filtered through the roots of the plant is not near so full freighted with nutrient particles as the chyle which enters the lacteals of

the animal; the plant, therefore, requires a much larger supply of fluid."

859. As soon in the animal series as organs are formed distinct from the homogeneous mass of which the minute and simple beings placed at the bottom of the scale appear to consist, these organs are appropriated, at least in part, to the function of excretion. In the human being, six organs take a part, and are chiefly appropriated to this function—namely, the skin, the lungs, the liver, the adipose tissue, the kidneys, and the intestinal canal. All these organs serve other purposes in the economy; but still the removal, in some specific form, of excrementitious matter from the system, is a most important part of the office of each.

860. The skin (34), to which are assigned numerous and highly important offices, seems to be specially constructed for performing the function of excretion. It is composed of three layers, of which the internal is called the cutis, or true skin; the external the cuticle, or scarf skin; and the middle, by which the other two are united, the rete mucosum. The latter is indistinct, excepting in the negro, in whom it is the seat of colour.

861. The cutis, or true skin, is a dense membrane, composed of firm and strong fibres, interwoven like a felt. Its internal surface is marked by numerous depressions, which receive processes of the adipose tissue beneath. Over its external

surface is spread a delicate and complex network of vessels, termed the vascular plexus, of such extent and capacity that, in the natural state of the circulation, a very large proportion of the whole blood of the body is constantly flowing in these blood-vessels of the cutis. A prodigious number of nerves accompany the cutaneous blood-vessels, some derived from the organic, and others from the sentient portion of the nervous system. The organic nerves endow the arteries with the power of performing the organic processes proper to the cutis, which are principally of an excrementitious nature. The sentient nerves communicate to every point of the external surface of the cutis the exquisite degree of sensibility possessed by the skin. Innumerable absorbent vessels terminate at the same points, with the capillary arteries and the sentient nerves.

862. The extreme smoothness and softness natural to the skin is communicated to it by a number of follicles which are placed in the cutis, and are termed sebaceous, from the oily substance they secrete. It is the matter secreted by these organs which communicates to the animal body the odour peculiar to it, on which the scent depends.

863. In many parts the cutis is perforated obliquely by hairs, which spring from little bulbs beneath it, to which the growth of the hairs is

confined. The human hair, which is hollow, consists of fine tubes filled with an oily matter. This matter is either of a black, red, yellow, or pale colour, as the hair is black, red, yellow, or white.

864. The nails are products formed by the cutis, and are essentially the same as the cuticle.

865. By long-continued boiling the cutis is resolvable into gelatin, which by evaporation becomes glue, and by combining with tannin and the extractive of oak bark is converted into leather.

866. The third portion of the skin, the cuticle, is a thin, elastic membrane spread over the external surface of the cutis, from which it is easily detached, by the action of a blister in the living, and by the process of putrefaction in the dead body. It is without vessels and nerves, and consequently it is insensible and inorganic. It is formed as a secretion by the cutis, and is composed almost entirely of solid albumen. When any portion of it is removed, it is renewed with great rapidity. Since it is subject to constant waste from friction, and is much increased by pressure, as is manifest in the palms of the hands and the soles of the feet, its formation must be continual; yet even in the fœtus it is thicker in the parts where pressure is ultimately to be made than in the other parts of the body.

867. The cuticle is a sheath in which the body

is enclosed for the purpose of restraining the organic actions which take place at its surface, and for tempering the sentient impressions received there. For restraining the organic actions it is fitted by the cohesion of its parts, which is such as to receive and transmit any fluid very slowly, as is manifest from the dryness of its surface when it is raised in a blister, and from the extreme rapidity with which the cutis dries, until it becomes as hard as parchment, when the cuticle is removed from it in the dead body.

868. Diffused over every part and particle of the cutis is the seat of common sensation, that cognizance may be taken of the presence of external objects. Restricted to particular points, the tips of the fingers, is the seat of one of the special senses, that of touch. Had the nerves which communicate to this extended surface its acute sensibility been placed in direct contact with external bodies, intolerable pain would have been the result; but by covering this surface with an inorganic and insensible substance, yet so thin that it is a pellicle rather than a membrane, the organ of sense is shielded, while the delicacy of the sensation is not impaired. But the control of the organic process and the protection of the sentient nerve are not the only offices performed by the cuticle; it serves further to hide what it is undesirable to have constantly in view. All that is beautiful in the blood as an object of sense is

rendered visible through the cuticle, in the bright and rosy hue of health, at the same time that every process, the sight of which would excite anxiety or terror, is effectually concealed.

869. The skin, an organ of secretion, an organ of absorption, an organ of excretion, and an organ of sense, is thus the immediate seat of three organic processes and of one animal process.

870. The chief excretion performed by the skin, in the human body, is commonly known under the name of perspiration. The perspiration is either sensible or insensible. Sensible perspiration is the liquid commonly called the sweat. Insensible perspiration consists of a vapour which, under the ordinary circumstances in which the body is placed, is invisible. The invisible vapour is constantly exhaling; the visible liquid is only occasionally formed. The quantity of matter carried out of the system under the form of invisible vapour is much greater than that lost by the visible liquid.

871. That a quantity of matter is incessantly passing off from the surface of the skin, under the form of an invisible vapour, is proved by the following facts :—

1. If the hand and arm are enclosed in a glass jar, the inner surface of the glass soon becomes covered with moisture.

2. If the tip of the finger be held at about the

twelfth of an inch from a mirror, or any other highly polished surface, the surface rapidly becomes dimmed by the vapour which condenses upon it in small drops, and which disappear on the removal of the finger.

3. If the body be weighed at different periods, an accurate account being taken of the ingesta and the egesta, it is found to undergo a loss of weight sensibly greater than can be attributed to any of the visible discharges: this loss must be owing to the transmission of a quantity of matter out of the body, under the form of invisible vapour.

872. The matters excreted under the form of perspiration are separated from the blood by a true and proper secretion, like the other secretions of the body. The process by which this is effected is called transudation. The matter of transudation deposited on the surface of the skin by a vital function is removed from the body by evaporation, a physical process which consists of the conversion of a liquid into a vapour by the addition of heat. Consequently the process of perspiration is a cooling process, and it is chiefly by the increase of the perspiration that the body is enabled to bear the intense degrees of heat which it has been shown (491, *et seq.*) to be capable of sustaining. Sitting one day in repose in the shade during the intense heat of an American summer's day, the skin freely perspiring at

every pore, Dr. Franklin happened to examine the temperature of his body with a thermometer. He found that the temperature of his body was several degrees lower than that of the surrounding air. The physiologists who exposed themselves in heated chambers, for the sake of ascertaining the greatest degree of heat which the human body is capable of enduring, perspired profusely during the experiment (495). The artisans who carry on their daily occupations in elevated temperatures perspire most profusely (884, *et seq.*). Under such circumstances, caloric is communicated to the human body just as freely as to inorganic matter yet it does not injure the body, because it does not accumulate in the system, but is immediately expended in supplying the heat necessary to convert the water, which is poured out upon the skin, into vapour. In this manner that surface of the body at which, under ordinary circumstances, a large portion of its animal heat is generated, is the very surface at which, under extraordinary circumstances, cold is generated, and the heat of the system positively reduced.

873. The physical process of evaporation would go on to a certain extent, though the vital function of transudation did not exist, and does go on in the dead body when the vital function is at an end. An organic tissue enclosing a liquid may not be porous enough to give passage to a single drop of liquid, and yet sufficiently porous to

admit air. In this case the air in contact with the tissue dissolves the liquid in its interior, and carries it off in the form of invisible vapour; hence liquids contained in organic bodies in contact with the air diminish in quantity by evaporation. But if an animal be placed in air saturated with moisture, and of the same temperature as its own, the air can no longer deprive that animal of a single particle of its moisture: evaporation from the body, in such a condition of the air, is suppressed. On the other hand, when an animal is placed in air saturated with moisture, and of the same temperature as its own, so far is transudation from being suppressed, that the sweat streams from every part of the external surface of the body. By modifying the condition of the air, in regard to its hygrometrical state and its temperature, the result of the physical process and of the vital function may thus be separated from each other, and the amount of each may be ascertained with perfect exactness. Now, by numerous experiments on the cold-blooded vertebrata, placed under such conditions of the air, it is found that, in these animals, perspiration by evaporation is to that by transudation as 6 to 1. But since the human body presents to the air an immense extent of surface over which is constantly flowing a large proportion of the whole quantity of blood contained in the system, the loss by the physical process compared with that by the vital function must be still greater in man than in the cold-blooded animal.

874. Taking together the average quantity of matter removed from the human body by both processes, or the whole loss of weight sustained from perspiration, on the comparison of the results of many observations, it is estimated to vary from twenty ounces in the twenty-four hours of the colder, to forty ounces in the warmer climates of Europe. Keill estimated it at thirty-one ounces. In the climate of Paris it is stated to be thirty ounces.

875. By the delicate tests of modern chemistry, various substances are found to be contained in the aqueous fluid which constitutes the great proportion of the matter of perspiration, namely, an acid, probably the lactic, a small proportion of animal matter, some alkaline and earthy salts, an oily or fatty substance, probably derived from the sebaceous follicles. All these matters are so analogous to the constituents of the serum of the blood as to leave little ground for doubt that they are merely separated from this part of the blood as it is flowing through the complex network of vessels spread over the surface of the cutis (861).

876. The skin, when in contact with the air, also separates a portion of carbon from the blood, and to the extent in which it does this

it is auxiliary to the lungs; but the quantity of carbonic acid excreted by the skin is small and variable in amount. The primary office of the skin as an organ of excretion is to relieve the blood of its superabundant watery particles, that is, to remove from the system its superfluous hydrogen.

877. A full account has been given (359, *et seq.*) of the primary office of the lungs, which, it has been shown, is to decarbonize the blood. The details of the calculations have been stated (457), from which it is estimated that 10 ounces and 116 grains of carbon are daily exhaled by the lungs under the form of carbonic acid; and the reasons have been assigned which favour the conclusion that the carbonic acid expired is not formed immediately in the lungs by the combination of the oxygen of the atmospheric air with the carbon of the blood; but in the system, where the oxygen taken into the blood at the lungs unites with carbon, the carbonic acid resulting from the combination passing as soon as formed into the capillary veins. The blood contained in these vessels, thus become venous, returns to the lungs, where it gives off the carbonic acid accumulated in it, and by that depuration again assumes its arterial character.

878. Some interesting experiments performed by Dr. Stevens appear to show that there exists a powerful attraction between oxygen and carbonic acid, and that the venous blood, as it is flowing

through the lungs, is freed from its carbonic acid by virtue of that attraction. Chemists were so universally agreed that the carbon in carbonic acid is united with its maximum dose of oxygen, that the idea of an attraction between carbonic acid and oxygen appeared highly improbable. The evidence of the fact, however, is decisive. If a receiver, filled with carbonic acid, and closed by a piece of bladder, firmly tied over it, be exposed to the atmospheric air, the carbonic acid, notwithstanding its superior specific gravity, rapidly escapes, and does so without the exchange of an equivalent portion of atmospheric air; the bladder is consequently forcibly depressed into the receiver. If the converse of this experiment be tried, and the receiver, containing atmospheric air, be tied over with a piece of bladder or thin leather, and then be immersed in carbonic acid, this gas will so abundantly penetrate the membrane and enter the receiver as to endanger its bursting.

879. Dr. Stevens had repeated opportunities of verifying these facts, during a stay which he made at Saratoga, in the United States, the springs at which place liberate a large quantity of carbonic acid. In the high rocks it often collects in considerable quantity and purity, and experiments on dogs and rabbits are often made for the entertainment of strangers, as at the Grotto del Cano, near Naples. This rock stands by itself in a low valley, through which there run two currents of

water, the one fresh and superficial, the other beneath and charged with salts and carbonic acid. A current of this water rises to some height in a cavity of the high rock, which appears to have been formed by a deposition of earthy salts from the water. It has a conical figure, the base of which is below the surface of the ground, and is about nine feet in diameter. It rises about five feet from the ground, where it is truncated, and presents an aperture a foot in diameter. The water rises in general only about two feet above the ground, and in the three feet above the surface of the water the liberated carbonic acid collects. By luting a large funnel over the aperture, carbonic acid may be collected at the mouth of the funnel in indefinite quantities, of which Dr. Stevens availed himself to multiply and vary his experiments, the result of which appears to be, the complete establishment of the fact that there exists a powerful attraction between carbonic acid and oxygen.

880. The application of this fact to the explanation of the phenomena of respiration is highly interesting. By virtue of this mutual attraction, two currents are established, which flow in opposite directions, through the membranous matter of the air-vesicles of the lungs and the pulmonary blood-vessels spread out upon their surface; the oxygen of the air flows to the blood attracted by its carbonic acid, and the carbonic acid of the blood

flows to the air attracted by its oxygen. According to Dr. Stevens, the moment the blood parts with its carbonic acid it loses its dark colour, and becomes of a bright vermilion colour, for the following reason: all acids impart a dark colour to the blood. With respect to most acids, this colour remains, although the added acid be afterwards saturated. Carbonic acid forms an exception, for on the removal of this aerial acid the blood resumes its bright and arterial colour. Alkalies, like acids, darken the colour of the blood, but salts produce a bright and vermilion colour when added to the colouring matter of the blood. When the blood loses its carbonic acid, the salts contained in the blood produce upon its colouring matter the vermilion tint natural to the combination when the influence of the salts is not counteracted by the presence of a redundant acid. At the moment the venous blood gives up its carbonic acid it receives in exchange a portion of the inspired air, which is chiefly at the expense of the oxygen. It retains somewhat more oxygen than it yields back in the shape of carbonic acid. The reddened and oxygenated blood, having returned to the heart, is diffused over the system, where it parts with its oxygen and combines with carbon, forming by the union carbonic acid; the necessary result of this combination is the generation of animal heat in the exact proportion to the quantity of the carbonic acid which is produced. The venous blood, which receives the carbonic acid as it is formed in the system, is darkened by its presence, which counteracts the effects of the salts of the blood upon its colouring matter.

881. An account has been given (439) of the experiments, which prove that the lungs also constantly exhale a quantity of azote.

882. It has been further shown (469) that, together with the carbonic acid, which passes off in the inspired air, there is always present a quantity of aqueous vapour. This aqueous vapour is not visible at the ordinary temperature of the air in its ordinary hygrometric state, because the water is then dissolved in the air, and is carried off in the form of invisible vapour; but it becomes abundantly manifest at a low temperature, or when the air is loaded with moisture. By the removal of this aqueous vapour, the lungs assist the skin in the depuration of the blood. The water transpired by the lungs, like that perspired through the skin, is separated from the blood by a true and proper secretion constituting the pulmonary transudation. It is commonly estimated that the lungs exhale about one-third as much as the skin, or fifteen ounces daily. Dalton estimates it at twenty-four ounces.

883. These estimates of the quantity of fluid lost by cutaneous and pulmonary transpiration relate to the quantities lost at the ordinary external temperatures in which the human body is placed

The quantity lost when the body is exposed to an elevated temperature is prodigiously increased. It did not occur to the physiologists, whose experiments have been detailed (492, *et seq.*), to ascertain this by causing themselves to be accurately weighed immediately before they entered their heated chamber and immediately after they left it. Having heard that the loss daily sustained by the workmen employed in gas-works is very extraordinary, I endeavoured to ascertain the amount of it with exactness. This I have been enabled to accomplish by the assistance of Mr. Monro, the manager of the Phœnix Gas Works, and of Mr. Cooper. The following are the experiments by which this has been ascertained.

EXPERIMENT I.—November 18, 1836, at the Phœnix Gas Works, Bankside, London.

884. Eight of the workmen regularly employed at this establishment in drawing and charging the retorts and in making up the fires, which labour they perform twice every day, commonly for the space of one hour, were accurately weighed in their clothes immediately before they began and after they had finished their work. On this occasion they continued at their work exactly three-quarters of an hour. In the interval between the first and second weighing, the men were allowed to partake of no solid or liquid, nor to part with either. The day was bright and clear, with much

wind. The men worked in the open air, the temperature of which was 60° Farh. The barometer 29° 25′ to 29° 4′.

	Weight of the Men before they began their work.				Weight of the Men after they had finished their work.				Loss.	
	cwt.	qr.	lbs.	oz.	cwt.	qr.	lbs.	oz.	lbs.	oz.
Michael Griffiths	1	1	14	10	1	1	12	2	2	8
John Kenny	1	0	26	10	1	0	24	1	2	9
John Ives...........	1	0	14	2	1	0	11	8	2	10
James Finnigan	1	1	10	6	1	1	7	0	3	6
William Hummerson	1	0	24	4	1	0	20	8	3	12
Timothy Frawley....	1	1	8	10	1	1	4	12	3	14
Patrick Nearey	1	1	14	10	1	1	10	8	4	2
Bryan Glynon	1	1	0	4	1	0	24	1	4	3

EXPERIMENT II.—Nov. 25, 1836.

885. Day foggy, with scarcely any wind. Temperature of the air 39° Farh., barometer 29° 8′. On this occasion the men continued at their labour one hour and a quarter.

	Before.				After.				Loss.	
	cwt.	qr.	lbs.	oz.	cwt.	qr.	lbs.	oz.	lbs.	oz.
Patrick Murphy	1	1	0	0	1	0	27	2	0	14
John Broderick	1	0	9	4	1	0	8	0	1	4
Michael Macarthy...	1	0	11	9	1	0	10	3	1	6
Michael Griffiths	1	1	15	8	1	1	13	2	2	6
James Finnigan.....	1	1	12	4	1	1	9	12	2	8
Bryan Duffy	1	1	11	12	1	1	9	0	2	12
John Didderick	1	1	11	5	1	1	8	8	2	13
Charles Cahell	1	1	4	5	1	1	1	6	2	15

886. Charles Cahell, the man who on this occasion lost the most, was weighed previously to the commencement of his work, with all his clothes off, excepting his shirt, which was kept dry and put on him again when weighed a second time at the end of his work. He was then immediately put into a warm bath at 95° Farh., and

kept there half an hour: he complained of being weak and faint, and when reweighed had gained half a pound.

EXPERIMENT III.—June 4, 1837.

887. Day clear, with some wind. Temperature 60° 5'.

	Before.				After.				Loss.	
	cwt.	qr.	lbs	oz.	cwt.	qr.	lbs.	oz.	lbs.	oz.
Robert Bowers	1	1	19	0	1	1	17	0	2	0
William Mullins	1	1	3	0	1	1	1	0	2	0
Charles Cahell	1	1	2	0	1	1	0	0	2	0
John Kenny	1	0	22	2	1	0	19	8	2	10
Bryan Glynon	1	0	27	0	1	0	24	4	2	12
John Haley	1	1	4	0	1	1	1	4	2	12
Benjamin Faulkner	1	1	15	14	1	1	13	0	2	14
Michael Griffiths	1	1	8	8	1	1	5	8	3	0
John Broderick	1	0	4	6	0	3	27	8	4	14
John Didderick	1	1	6	12	1	1	1	10	5	2

888. The two last men worked in a very hot place for one hour and ten minutes; all the rest worked about one hour. Michael Griffiths, as soon as he had finished his work, was put into a bath at 98°, where he remained half an hour. He was reweighed on coming out of the bath, and had lost 8 oz.

889. From these observations it appears that, towards the end of November, when the temperature of the external air was 39°, and the day was foggy and without wind, the greatest loss did not amount to 3 lbs. (2 lbs. 15 oz.), the least loss was 14 oz., and the average loss was 2 lbs. 3 oz.

890. In the middle of the same month, when

the temperature of the air was 60°, and the day was clear with much wind; the greatest loss was 4 lbs. 3 oz., the least loss was 2 lbs. 8 oz., and the average loss was 3 lbs. 6 oz.

891. In June, when the temperature of the external air was 60°, and the day exceedingly bright and clear, without much wind, the greatest loss was 5 lbs. 2 oz.; the next greatest loss was 4 lbs. 14 oz., the least loss was 2 lbs., and the average loss was 2 lbs. 8 oz.

892. The same individuals lose very different quantities at different times. Thus, James Finnigan in the first experiment lost 3 lbs. 6 oz., in the second 2 lbs. 8 oz. Michael Griffiths in the first experiment lost 2 lbs. 8 oz.; in the second 2 lbs. 6 oz., and in the third 3 lbs.; while John Kenny in the first experiment lost 2 lbs. 9 oz., and in the third experiment, which was the second to which he was subjected, he lost very nearly the same, namely, 2 lbs. 10 oz. On the other hand, Bryan Glynon in the first experiment lost 4 lbs. 3 oz., and in the third experiment, which was the second to which he was subjected, he lost no more than 2 lbs. 12 oz.

893. In one case, when a man who had lost 2 lbs. 15 oz., the greatest quantity lost by any of the men examined during that day, was put into a hot bath at 95°, and reweighed on coming out of the bath, where he had remained exactly half an hour, it was found that he had gained half a

pound. On the other hand, when a man who had lost 3 lbs. was put into a hot bath at 98°, and kept there for half an hour and reweighed, it was found that he had lost exactly half a pound.

894. It was our intention to have pursued these experiments, with the view of ascertaining the influence of the hygrometrical state of the air on transpiration, as well as the absorbing power of the skin, under circumstances so favourable to the activity of that power, but the investigation has been unavoidably postponed.

895. The results of these observations are as interesting in relation to absorption as to transpiration. Thus, James Finnigan, on the 18th of November, weighed,

	cwt.	qr.	lbs.	oz.
before the experiment	1	1	10	6
after the experiment . .	1	1	7	0
having lost	0	0	3	6

On the 25th of November he weighed 1 cwt. 1 qr. 12 lbs. 4 oz., having gained in the interval 1 lb. 14 oz.

Michael Griffiths, on the 18th of November,

	cwt.	qr.	lbs.	oz.
before the experiment, weighed	1	1	14	10
after the experiment	1	1	12	2
having lost	0	0	2	8

On the 25th of November, before the experiment, he weighed 1 cwt. 1 qr. 15 lbs. 8 oz., having gained 14 oz.; but on the 3rd of June he weighed 1 cwt. 1 qr. 8 lbs. 8 oz., having lost between the 18th of November and the 3rd of June, 6 lbs. 2 oz.

896. John Kenny, on the 18th of November,

	cwt.	lbs.	oz.
before the experiment, weighed .	1	26	10
after the experiment	1	24	1
having lost	0	2	9

On June the 3rd he weighed 1 cwt. 22 lbs. 2 oz., having gained in the interval 4 lbs. 8 oz.

897. Bryan Glynon, November 18th,

	cwt.	qr.	lbs.	oz.
before the experiment, weighed	1	1	0	4
after the experiment	1	0	24	1
having lost	0	0	4	3

On the 3rd of June he weighed 1 cwt. 27 lbs., having lost 1 lb. 4 oz.

898. Thus, in the course of their ordinary occupation, these men are in the habit of losing from 2 lbs. to 5 lbs. and upwards twice a-day; yet, when weighed at distant intervals, it is found that some have actually gained in weight and others have lost only a few pounds; it follows that the activity of the daily absorption must be proportionate to that of the daily transpiration.

899. According to the prevalent opinion, the liver is the cause of a large proportion of the maladies which afflict and destroy human life It certainly

exercises an important influence over health and disease, the true reason of which is but little understood by those who attribute most to its agency.

900. The liver is an organ of digestion and an organ of excretion.

It is an organ of digestion in a two-fold mode:

1. By the secretion of a peculiar fluid, through the direct action of which chyme is converted into chyle. The several phenomena attending this operation have been fully described (668 *et seq.*).

2. By subjecting alimentary matters which have been partly acted on by the stomach and intestines to a second digestion.

901. It has been shown (666) that the veins which return the blood from the digestive organs, the stomach, the intestines, and the mesentery, together with the veins of the spleen, the omentum and the pancreas, instead of pursuing a direct course to the right side of the heart in order to transmit their contents by the shortest route to the lungs, as is the case with all the other veins of the body, unite together and form a large trunk termed the vena portæ, which enters the liver and ramifies through it in the manner of an artery. It has been further shown (666) that the bile is secreted from the venous blood contained in this vessel by its capillary branches spread out on the walls of the biliary ducts, the only known instance in the human body in which

a secretion is formed from venous blood by venous capillaries; that the trunk of this vein, unlike that of any other, is encompassed with organic nerves, which accompany its subdivisions, and are spread out upon its capillary branches just as an organic nerve is spent upon an artery, and that thus, as this vessel performs the function of an artery, it has the structure and distribution of an artery.

902. The veins which unite to form the vena portæ take up, by their capillary branches, certain portions of the contents of their respective organs, and bear those contents directly into the venous current. The capillary veins of the stomach take up certain parts of the contents of the stomach, it would appear the fluid substances received with the aliment more especially; the capillary veins of the duodenum take up certain portions of the contents of the duodenum, and so on of the capillary veins of the spleen, intestines, and all the organs whose veins combine to form the vena portæ. Further, branches of the absorbent vessels of these organs have been distinctly traced opening directly into the veins in their immediate neighbourhood. Certain products of digestion must, then, be constantly poured, both by the capillary veins and by the absorbent vessels of the digestive organs, into the blood of the vena portæ.

903. Accordingly, on the examination of animals soon after a meal, streaks of a substance like

chyle are often observed in the blood of the vena portæ. It is further established by numerous experiments, that if alcohol, gamboge, indigo, and other odoriferous and colouring matters, are mixed with the food, their presence is manifest in the blood of the digestive organs, and more especially in the blood of the mesenteric veins and in that of the vena portæ, while no trace of these substances is ever found in the lacteals.

904. The lacteals, it has been shown (835. 1), are special organs appropriated to the performance of a specific function, that of absorbing chyle. To fit them for this office, they are endowed with an elective power, by virtue of which they select, from the alimentary mass, that portion of it only which is converted into chyle; in a natural and healthy state they would appear to be incapable of absorbing any other substance excepting pure chyle. But in the digestive organs there is always present much nutritive matter not yet converted into proper chyle, and with this matter there are mixed foreign substances not strictly alimentary. These unassimilated matters and foreign substances, absorbed by the capillary veins or by the absorbent vessels, or by both, are conveyed directly into the vena portæ, by which vessel they are transmitted to the liver, where they undergo a true and proper digestion. After undergoing this digestion in the liver, they are sent by a short course to the heart, and thence to the lungs, where they are assimilated

into, or at least commingled with, arterial blood, and, with arterial blood, are transmitted to the system. The substances subjected to this hepatic digestion, which is as real as that effected in the stomach and duodenum, do not appear to enter the lacteals at all; they have therefore a shorter course to traverse, and probably a proportionately less elaborate process to undergo, before their transmission to the lungs and their final entrance into the arterial system.

905. What the particular substances are for which this slighter digestive process suffices is not known with certainty. There is, however, reason to suppose that they consist chiefly of liquids, while there is direct evidence that vinous and spirituous liquids enter the system through this shorter course; since these fluids are often abundantly manifest in the blood of the vena portæ, when not the slightest trace of them can be detected in the lacteal vessels.

906. According to this view, the liver is a second digestive apparatus, completing what the first commences, or effecting what that is incapable of accomplishing; and this view assigns the reason why certain fluids taken into the stomach sometimes appear in the secretions and excretions with such astonishing rapidity; why the liver so constantly becomes diseased when highly stimulating substances, not properly alimentary, are mixed with the food, and more especially when

ardent spirits or the stronger wines are largely and habitually taken; why the sympathy is so intimate and intense between the stomach and the liver and the liver and the stomach, both in health and disease; why in the ascending animal series the liver so soon appears after the stomach, and why the magnitude of the organ and the elaborateness of its structure progressively increase with the extension of the digestive apparatus and the corresponding complexity of the general organization.

907. The second function performed by the liver is that of excretion. The excrementitious matter eliminated from the blood by the liver is contained in its peculiar secretion, the bile. The bile consists of two portions, an assimilative part which combines chemically with the chyle, purifying and exalting its nature; and an excrementitious part which combines with the residue of the aliment.

908. The excrementitious part of the bile contains a large proportion of carbon and hydrogen. Carbon and hydrogen abound in venous blood; venous blood in large quantity is sent to the liver to afford the materials for the secretion of bile; consequently, the more copious the secretion of bile the greater the quantity of carbon and hydrogen abstracted from venous blood. It follows that, by this elimination of carbon and hydrogen

from the blood, the liver is auxiliary, as an organ of excretion, to the skin and the lungs.

909. But it is well worthy of remark, that although the liver at all times assists the skin and the lungs in carrying on the process of excretion, it does this most especially under circumstances which necessarily enfeeble the action of the cutaneous and pulmonary organs.

910. Less carbon is expelled from the lungs in summer than in winter; at a high than at a low temperature; consequently by a long-continued exposure to intense heat, as in the hot months of summer, and still more by a continual residence in a warm climate, an accumulation of carbon in the blood is favoured. A part of this excess is removed by the increased exhalation from the skin. The skin, however, is the chief outlet, not for carbon, but for hydrogen; and accordingly by the increased perspiration hydrogen is largely removed. Hydrogen and carbon compose fat. The deposition of fat, could it go on to the requisite extent, would afford an adequate consumption for the superabundant carbon; but the formation of fat is prevented by the dissipation of the hydrogen. Under such circumstances, when the lungs cannot carry off the requisite quantity of carbon, nor the adipose tissue compensate for its diminished activity by the deposition of fat, the liver, taking on an increased action, secretes an

extraordinary quantity of bile. In this manner the superfluous carbon, instead of being removed in the ordinary mode, by the pulmonary artery through the lungs, under the form of carbonic acid gas, is excreted by the vena portæ, through the liver, under the form of bile, while the superabundant hydrogen is removed by the increased quantity of perspiration; and thus the accumulation of these inflammable matters in the system is effectually prevented.

911. By the deposition of fat in the adipose tissue material assistance is afforded to the excretory action of the skin, the lungs, and the liver. Fat is composed essentially of carbon and hydrogen; it contains no nitrogen and very little oxygen. It is deposited whenever an excessive quantity of nutritive matter is poured into the blood, and especially when at the same time the different secretions and excretions ordinarily formed from the blood are diminished. The primary object of this deposition is to relieve the circulation of a load which would embarrass and ultimately stop the actions of life. It serves, however, a secondary purpose, that of forming a storehouse of nutritive matter, duly prepared for supplying the wants of the system, in case the body should be placed under circumstances in which the digestive organs can no longer receive food or no longer convert it into chyle.

912. Thus hybernating animals, which pass

many months without taking food, accumulate a store of fat before they fall into the state of torpor. Marmots and dormice subsist on this store during the winter, and hence, when spring awakens them from their torpor, they are always in a state of extreme emaciation. Birds and other animals which live on food procured with difficulty in the winter, become unusually fat in the autumn.

913. During fever and other acute diseases, when little food is received, and still less converted into chyle, the extreme emaciation which the body undergoes is owing partly to the disappearance of the fat, which is taken up by the absorbents and carried into the blood, in order to compensate for the deficiency of nutrient matter supplied by the digestive organs.

914. The chief depositories of the fat are those intersticial spaces of the body in which a certain quantity of soft but tenaceous substance is required to obviate pressure or to preserve symmetry. A large quantity is also placed immediately beneath the skin; in the interstices of muscles; along the course of blood-vessels and nerves; in the omentum, where it is spread like a covering over the viscera of the abdomen (fig. CLXX. 7); in the mesentery and around the kidneys.

915. Fat is a bad conductor of heat; consequently the layer which is spread over the external surface immediately beneath the skin, and

that which is collected in the interior of the omentum, must be useful in preserving the heat of the body. Fat persons bear cold better than lean persons. Animals which inhabit the northern climates, and the fishes of the frozen seas, are enveloped in prodigious quantities of fat. Where the accumulation of this substance would produce deformity or interfere with function, as about the joints, in the eyelids, within the skull, not a particle is ever deposited. About the joints it would impede motion ; in the eyelids it would render the face hideous and obstruet vision ; and within the skull, a cavity completely filled with the brain, an organ impatient of the slightest pressure, had a substance been placed, the quantity of which is liable to be suddenly trebled or quadrupled, changes in the system which now produce no inconvenience would have been fatal. Thus, while provision is made at once to exonerate the system from too great a load of nourishment, and to lay up the superfluous matter, as in a magazine, to be ready for future use, the most extreme care is taken to deposit the store in safe and convenient situations.

916. The excretory organs and processes, hitherto considered, have for their object the removal from the blood of its superfluous carbon and hydrogen ; the element peculiar to the animal body, azote, is eliminated by the kidneys, glan-

dular organs which possess a highly complex structure.

917. But besides the removal of the superfluous azote, the fluid secreted by the kidneys would appear to be a general outlet for whatever is not required in the system, and for the removal of which no specific apparatus is provided. Chemical analysis shows that, in different states of the system, the following substances are contained in this fluid :—water, free phosphoric acid, phosphate of lime, phosphate of magnesia, floric acid, uric acid, benzoic acid, lactic acid, urea, gelatin, albumen; lactate of ammonia, sulphate of potash, sulphate of soda, fluate of lime, muriate of soda, phosphate of soda; phosphate of ammonia, sulphur, and silex.

918. This catalogue itself suggests the idea that when any matter employed in carrying on the functions is in excess, or when it has become decayed, or is decomposed and is not eliminated by any other excretory process, it is taken up by the absorbents, poured into the veins, and so conveyed in the course of the circulation to the kidneys, by which organs it is separated from the blood, and thence by an appropriate apparatus carried out of the system.

919. The specific matter secreted by the kidneys is that termed urea; a substance of a resinous nature, highly animalized. One charac-

ter by which the animal is distinguished from the plant is its locomotion. The organ by which the animal is rendered capable of performing the function of locomotion is muscle or flesh. The basis of muscle is fibrin, and the basis of fibrin azote. There must be in the animal body an abundant supply of fibrin, and consequently a proportionate abundance of azote. Azote is introduced into the system partly by the food and partly by the lungs. That there may be a sufficiency for all occasions, more is introduced than is necessary on ordinary occasions, and a special outlet is established for the excess through the kidneys.

920. Organs appropriated to the removal of substances from the blood, capable of becoming deleterious by their accumulation, generally in a state of health perform their office so perfectly that the matters which it is their part to excrete are eliminated almost as quickly as they enter the blood, so that they are seldom present in the circulating fluid in sufficient quantity to be detected by the most delicate chemical tests. But by the removal of the excretory organ, or by the suppression of its function, the excretory matter accumulates in the blood, and is then readily detected. A decisive experiment disclosed that this is the case with regard to urea. The kidneys were removed from a living animal. The operation did not appear to be productive of materia'

injury for some time; but at length symptoms denoting the presence of a poison in the blood arose, and the animal died. The blood was carefully examined after death. It was found to contain a much larger quantity than ordinary of the peculiar animal substance which enters into the composition of the serosity of the blood (225). On subjecting this substance to the action of various re-agents, and also on reducing it to its ultimate elements, it was found to resemble urea; to be, in fact, nearly identical with urea as contained in the urine. From this experiment it became manifest that the source of the urea is the serosity of the blood. It is probable that the chief office of the kidney is to separate the urea from the other ingredients of the blood, and to convey it to the organs which are destined to carry it out of the body.

921. It is estimated that about a thousand ounces of blood pass through the kidneys in the space of an hour; itself a sufficient indication of the importance of the excretion performed by this organ, and an adequate source of the matter actually excreted, although, under ordinary circumstances, distributed through the circulating mass in quantities so minute as to be almost inappreciable.

922. From the power of absorption possessed by the veins of the stomach and intestines, from the connexion proved to be established between

the venous and absorbent systems, and from the discovery of Lippi, that several absorbent branches in the abdomen terminate directly in the pelvis of the kidney, that is now an established fact which was long a conjecture, that there exists a short route from the stomach to the kidneys, so that the extreme rapidity with which certain substances mixed with the aliment appear in the fluid secreted by the kidneys is no longer a matter of wonder.

923. Out of the body urea putrifies with great rapidity. When retained in the system by the extirpation of the kidney, or by placing a ligature around the ureter, such is the septic tendency communicated to the blood that signs of putrescency become manifest even during life, and after death all the soft parts of the body are reduced to a state of putrefaction with extreme rapidity. The suppression of the secretion in the human body, or the undue retention of the matter secreted, induces fever of a malignant kind, in which the symptoms that denote a highly putrid taint in the system are rapidly developed. But for the labour of the kidney, then, a substance would accumulate in the blood, which would quickly lead to the decomposition of the body.

924. It has been shown that the mucous membrane which lines the alimentary canal is studded in its whole extent with glands, which secrete from the blood a large quantity of fluid,

These secretions go on without interruption, whether food be taken or not, so that there may be copious alvine evacuations though not a particle of food enter the stomach; and the separation of the matter eliminated from the blood by this extended membrane can no more be dispensed with than that by the skin or the lungs. There is, too, a most intimate sympathy between the secretion of the membrane that lines the internal surface of the body and that carried on by its external covering; any disorder of the one immediately and powerfully disturbs the natural course of the other: hence the diarrhœa, so often produced by the application of cold to the external skin, and the diseases of the skin, so constantly connected with a disordered state of the mucous membrane of the intestines.

925. It is the special office of the large intestines to prepare for its removal, and to carry out of the system the residue of the aliment, together with the excrementitious portion of the bile.

926. It was calculated by Haller, that the different excretory organs remove from the system every twenty-four hours twenty pounds of matter. Of this total loss sustained daily by the human body, it was estimated that four pounds are removed by the skin, four pounds by the lungs, four pounds by the kidneys, and eight pounds by the intestinal canal. In this estimate, which is considered too

large, especially that by the intestinal canal, the quantity stated must be understood as denoting the maximum of each secretion.

927. Supposing the ingesta in twenty-four hours to be of food 6 pounds, or 96 ounces, and of oxygen retained in the system 4 ounces, in all 100 ounces, it is estimated that the egesta will be, in twenty-four hours, by the skin, 34 ounces, by the lungs 17 ounces, by the intestines 6 ounces, by the kidneys 40 ounces, and by various other excretions 3 ounces, in all, 100 ounces. These calculations must of course be taken only as approximations to the truth, and as ascribing rather the relative than the positive quantities of matter excreted.

928. Whatever be the absolute quantity or the form of the excretions, it is clear, from the preceding account, that there is constantly removed from the system by the skin a large portion of hydrogen and some carbon; by the lungs a large portion of carbon and some hydrogen; by the liver a large portion of hydrogen and some carbon; by the kidneys a large portion of azote; by the large intestines the residue of the aliment; while, by the deposition of fat, the surperabundant nutriment withdrawn from the current of the circulation is laid up in store in some safe part of the body.

929. Most of the processes which have been described are mutually compensating and vicarious.

Besides the office which each habitually performs, it is capable of having its action occasionally increased, for the purpose of supplying the deficiency of one or more of its fellows. If perspiration by the skin languish, transudation by the lungs increases; if neither the skin nor the lungs be able to remove the superfluous hydrogen and carbon, these inflammable substances are carried out of the system by the liver in an augmented secretion of bile. If the action of the liver be diminished, that of the kidney is increased; and if the secretion of urine be suppressed, the secretion of bile is augmented. When the absorbents are oppressed by the quantity of fluid poured into the stomach, or when the system is at the point of saturation, and no absorption can go on, the veins take up the superfluous liquids, pour them into the circulating current, and bear them to the kidneys, by which organs they are rapidly separated from the blood, and carried out of the body. The weakness of one organ is compensated by the strength of another; the diminished activity of one process is equalized by the increased energy of some other to which it is allied in nature and linked by sympathy; and thus the evils which would result from the partial and temporary failure of an important function are obviated by some vicarious labour, until the enfeebled organ has recovered its tone, and the natural balance of the functions is restored.

930. The condition acquired by the elementary

particles of organized bodies, from their long continuance in the system, which induces the necessity for their excretion, is not known. The chemical elements of the excretions are the very same as those which constitute the organized textures and the nourishment by which they are sustained. Carbon is the basis of the organized body; yet all living bodies, without exception, excrete carbon. Oxygen, hydrogen, and azote, also, without which life cannot be maintained, if retained in the system beyond a given time, are incompatible with the continuance of life. During the chemical changes which these elementary particles undergo, in the course of the vital processes, they appear to enter into some combination, which is no longer compatible with the peculiar mode in which they are disposed in organized and living structures. And one such change, of a very remarkable nature, has been observed, which, it is conceived, has a considerable share in rendering their constant expulsion and renovation indispensable.

931. Out of the condition of life the component elements of organized bodies readily combine so as to form crystals; the peculiar combinations by which they form the constituent textures of organic structures are never crystalline. No crystal is ever seen in the seat of a living and growing vegetable cellule; no crystal is ever found as a constituent part of animal membrane.

Whenever a crystal occurs in an organized body it is always the result either of disease or of some artificial process, or else it is an excretion separated from the nourishing fluid and the useful textures. Every one of these textures contains, even in its minutest parts, saline and earthy, as well as vegetable or animal, matter. Why do not these saline and earthy particles as readily combine to form crystals in the organic as they do in the inorganic body? They never do. In the organic body these saline and earthy particles are always so arranged that they are diffused through the membranous fibres or cells, never concentrated in crystals.

932. On the other hand, the elements containing the peculiar matters of excretion are generally in such a state of combination as readily to assume the crystalline form, either alone or in the simplest further combinations of which they are susceptible. It seems probable that this circumstance may be, at least in part, the cause which necessitates their expulsion, and it is certain that some such general principle must determine the incompatibility of the matters of excretion with the life of the structures

933. The ultimate object of the processes included in the function of excretion is to maintain the nutritive fluid in a certain chemical condition. Into the combination of the blood there must enter certain constituents, and these must be in certain relative proportions, and in no others. If

the salts be diminished or in excess, if the fibrin, or the red particles, or the serum be abundant or defective beyond a certain degree, either the necessary chemical elements are not present, or not present in the form necessary to their entering into the requisite combinations; the result is, that a proper nutritive fluid is not formed, and consequently due nourishment is not afforded to the textures nor due stimulus to the moving powers; there is either too much nutriment and stimulus or too little; in the one case the machine is exhausted and worn out, and in the other it is clogged and stopped.

934. The capillary arteries of the skin, and of all the other tissues into the composition of which gelatin enters as a constituent, necessarily pour carbon into the capillary veins at the moment they convert albumen into gelatin (539). The veins, receiving in their course more and more carbon from the arteries, at length become loaded with this element, and in order to get rid of the excess they bear it to the lungs, where it is expelled by the act of expiration under the form of carbonic acid gas. On the other hand the chyle, gradually becoming firmer and more condensed by the series of changes which it undergoes from its first formation in the duodenum to its admixture with the lymph in the receptacle of the chyle, and with the blood in the subclavian vein, is hurried to the heart and thence to the lungs, where it gives off a

large portion of its watery particles, also by the act of expiration, under the form of aqueous vapour. This excretion of its watery particles is a necessary part of the process of completion by which the weak albumen of the chyle is converted into the strong albumen of the blood (703. 3). How completely analogous then is this excretory process in the plant and in the animal! How precisely the same is the action of the leaf and of the lung! The leaf dissipates the superfluous water of the crude sap, concentrates its organic principles, and brings it into the chemical condition which constitutes the proper juice of the plant; the lung removes the superfluous water of the chyle, concentrates its organic principles, and completely assimilates its chemical nature into that of the blood.

935. It is the same with every other process of excretion; its uniform result is to alter the chemical composition of the nutritive fluid, to restore it to a state of concentration and purity. Excretion then is appropriately termed a depurating process.

936. The effect of the suppression of excretion, when the suppression is complete, is appalling. Stop the respiration, that is, suspend the depurating action of the lungs, carbon accumulates in the venous blood; carbon mixes with the arterial blood; in half a minute the blood flowing in the arteries is evidently darkened; in three-quarters of

a minute it is of a dusky hue; in a minute and a half it is quite black; every particle of arterial blood has now disappeared, and the whole mass is become venous. With the first appearance of the dusky hue great disturbance is produced in the system; the instant it becomes dark sensibility is abolished; in a few minutes after it is black the power of the heart is so enfeebled that it can no longer carry on the circulation, and in a few minutes more its action wholly ceases, and can never again be excited. The brain feels the poison first, and is first killed; but the heart cannot long resist the fatal influence.

937. Stop the excretion of the kidney by the extirpation of the organ, or the suppression of its secretion, urea accumulates in the blood; the poison, after a short time, begins to work; fever is excited, and then, with fearful rapidity, fever is followed by coma, and coma by death.

938. Stop the secretion of bile, a poison accumulates in the blood as potent, producing insensibility and death as rapidly, as that generated by the suppression of the depurating action of the kidneys.

939 Only obstruct the secretion of bile, merely prevent its due elimination from the blood, just in proportion to its suppression does the system suffer from languor, lassitude, and inaptitude for every muscular and mental exertion.

940. How do the internal organs suffer when

the excretion of the skin is deficient, and how numberless and hideous are the diseases of the skin when the depurating process of the alimentary canal is suspended!

941. When, on the contrary, all these excretions are well and duly performed, how regular and tranquil, yet how full and strong the flow of the circulating current; how rich the stream poured by it into every organ; how healthfully exciting its influence on them all; how gentle, how efficient, every organic action; how complete the absence of all note or sensible intimation that any such action is going on, yet how delicious the consciousness produced by its soundness and vigour; how acute the sense, how bounding the motion, how quick the percipience; how the pure blood mantles in the cheek and diffuses its sparkling colour over all the transparent complexion; how the jocund spirits laugh from the eyes; how the intellectual and sympathizing mind beams forth from them with a higher and holier happiness! How wonderfully beautiful is such a human body, and how magnificently endowed in its capacity to give and to receive enjoyment!

942. There are two adjustments, with regard to the excretions, carried on by organized bodies, which can never be contemplated with sufficient admiration. It has been fully shown (464 *et seq.*) that the relation established between the two great classes of organized beings is such that the ex-

crementitious matter of the plant is nutritious to the animal, and the excrementitious matter of the animal is nutritious to the plant; and, consequently, that the two orders of living beings maintain the world, which is given them as their inheritance, in a state of perpetual adaptation for the life and health of each other; the animal receiving healthy stimulation from that which is poisonous to the plant, and the plant being nourished by particles which the animal throws off as exhausted and useless. And this relation naturally suggests that so beautifully described by Milton:—

> Flow'rs and their fruit,
> Man's nourishment, by gradual scale sublimed
> To vital spirits aspire, to animal,
> To intellectual; give both life and sense,
> Fancy and understanding; whence the soul
> Reason receives.

943. Secondly, the particles thrown off by organized bodies are rendered, in the very act of their dissipation, subservient to purposes of utility and pleasure. How these poisonous elements are converted into the pabulum of life and health has been shown. To a being with the senses and faculties of man, how loathsome might these particles have been rendered during the period of their transition from one organized kingdom to the other! And if disagreeable at all, how constantly forced upon his sense, wherever he might be, during every moment of his waking hours, must these objects of disgust have been! But how

does the matter actually stand? The excretions of the plant are the very particles that, poured

> "Into the blissful field through groves of myrrh,
> And flow'ring odours, cassia, nard, and balm,"

create "a wilderness of sweets." It is as these exhalations are passing off from the economy to which, if retained, they would be noxious (851), that they become

> "Exhalations of all sweets
> That float o'er vale and upland;"

and which refresh and delight even more than the forms and colours of the "aery leaf" or "the bright consummate flower."

944. And the human body, when the functions of its economy are sound and vigorous, is fresh and fragrant as the flower (862); and by that intellectual faculty by which man is capable of associating his conception of beauty and delight with whatever object has been the source of exquisite gratification, the fragrance of the flower is but suggestive of what, to him, is inexpressibly sweeter and dearer.

> "As new waked from soundest sleep,
> Soft on the flow'ry herb I found me laid
> In balmy sweat, which with his beams the sun
> Soon dry'd——
> By quick instinctive motion up I sprung,
> ——— And upright
> Stood on my feet.——
> ——— All things smiled
> With fragrance, and with joy my h art o'erflow'd.
> Myself I then perused, and limb by limb
> Survey'd, and sometimes went, and sometimes ran,
> With supple joints, as lively vigour led. MILTON.

—— Fresh lily,
'Tis her breathing that
Perfumes her chamber thus. SHAKSPEARE.

—— The very air
With her sweet presence is impregnate richly,
As in a mead that's fresh with youngest green
Some fragrant shrub exhales——
Ambrosial odours——
Charming present sense,
And sure of memory;—so her person bears
A natural balm—distilling incense.

"Death of Marlowe," by R. H. HORNE.

CHAPTER XIV.

OF NUTRITION.

Composition of the blood—Liquor sanguinis—Recent account of the structure of the red particles—Formation of the red particles in the incubated egg—Primary motion of the blood—Vivifying influence of the red particles—Influence of arterial and venous blood on animal and organic life—Formation of human blood—Course of the new constituents of the blood to the lungs—Space of time required for the complete conversion of chyle into blood after its first transmission through the lungs—Distribution of blood to the capillaries when duly concentrated and purified—Changes wrought upon the blood while it is traversing the capillaries—Evidence of an interchange of particles between the blood and the tissues—Phenomena attending the interchange—Nutrition, what, and how distinguished from digestion—How the constituents of the blood escape from the circulation—Designation of the general power to which vital phenomena are referrible—Conjoint influence of the capillaries and absorbents in building up structure—Influence of the organic nerves on the process—Physical agent by which the organic nerves operate—Conclusion.

945. The object of the greater part of the processes hitherto described is to form the nutri-

tive fluid, and to bring it to the requisite state of purity and strength. Recent researches into the composition of the nutritive fluid confirm the general correctness of the account already given of it. (211 *et seq.*)

946. When examined as it is flowing in the finest vessels of a transparent part of the body, or immediately after it is abstracted from the trunk of a vein or artery, before coagulation (218) takes place, the blood is seen to consist of a colourless fluid, through which is diffused a countless number of minute solid particles of a red colour. The colourless fluid is called the liquor sanguinis, and the solid particles the blood corpuscles or the red particles.

947. By the process of coagulation, the phenomena of which have been fully described (219 *et seq.*), the blood spontaneously separates into a clear fluid of a yellow colour called serum or blood-water, and into a solid mass termed the clot or the crassamentum. The serum, which must be carefully distinguished from the liquor sanguinis, is the fluid formed from the blood by coagulation; the liquor sanguinis is the fluid part of the blood which exists before coagulation.

948. The liquor sanguinis contains in solution a large quantity of animal matter, fibrin (228), which separates spontaneously in a solid form on coagulation; the serum also contains a quantity of animal matter in solution, albumen (224), which

does not separate in a solid form spontaneously, but only on the application of heat, acids, alcohol, &c. (224.) The animal matter, the fibrin, which separates spontaneously from the liquor sanguinis in a solid form, constitutes one part of the clot, and the other part of it consists of the red particles which floated in the liquor sanguinis.

949. Thus, by coagulation, the liquor sanguinis separates into a portion which remains fluid, the serum; and into a portion which becomes solid, the fibrin; while the fibrin, as it is passing from the fluid to the solid state, entangles the red particles, and both together form the clot; consequently the liquor sanguinis contains in solution two kinds of solid matter, fibrin and albumen; while the serum contains in solution only one kind of solid matter, albumen.

950. The solution of fibrin in the liquor sanguinis, and its spontaneous solidification during the process of coagulation, has been shown by Professor Müller in the following mode. Having carefully collected blood from the femoral artery of the frog, and also from the heart laid bare and incised, and having brought a drop of this pure blood under the microscope, and diluted it with serum, so that the red particles were separated from each other by distant intervals, he observed that there formed in those intervals a coagulation of previously dissolved matter, by which the separated red particles were connected together. By

raising, with a needle, the coagulum occupying the intervening spaces, this solid matter was obtained free from red particles. The blood corpuscles of the frog are rendered, by a powerful microscope, so large, that this operation may be performed with the greatest distinctness. In consequence of the minuteness of the red particles of human blood they pass, with the liquor sanguinis, through filtering-paper; but those of the frog, being four times larger, are kept back by the filter, while the liquor sanguinis percolates through as a clear fluid, and then coagulates. This colourless coagulum is so transparent that it is not even detected, after its formation, until it is raised out of the fluid with a needle. It gradually thickens and becomes white. It is the fibrin of the blood in its purest state.

951. Professor Müller's account of the structure of the red particles differs in a material point from that given (231 *et seq.*). He agrees that they are rounded bodies (fig. CXII. 1), generally of the same size, though some are seen larger than common, but never double the mean diameter; that they are always quite flat (232); that in a certain light they look as if they were hollowed out from the edges to the centre (fig. CXII. 1); but, he adds, "that this spot is a real depression, as some think, appears to me in the highest degree improbable; for I have at last convinced myself that the blood corpuscles of

man and the mammalia contain a very small nucleus of the diameter of the flat corpuscle. My observations prove beyond doubt that the blood corpuscles of frogs and salamanders (fig. CXII. 4) contain a nucleus entirely different in its chemical relations from the outer layer. With one of Frauenhofer's microscopes I have seen very distinctly, in the blood corpuscles of man an exceedingly small, round, well-defined nucleus, yellower and brighter than the transparent circumference. When the blood corpuscles are mixed, under the microscope, with acetic acid, the shell is almost entirely dissolved, and these small nuclei, which are seen with great difficulty in human blood, remain, while those of the frog appear, very evidently the nuclei observed earlier in the blood corpuscles. In man, the nuclei within the corpuscles are so small, that the diameter does not exceed the thickness of the flat corpuscles."

952. The enveloping capsule is stated to be soluble in water, while the internal nucleus is insoluble; but the capsule is not soluble in serum; the albumen and the salts contained in the serum probably rendering it insoluble. The colouring matter of the capsule, which gives the red colour to the blood, is called hæmatosin. Lecanu considers the capsular substance as a combination of a specific colouring matter, which he calls globulin, and of albumen; but Müller regards it as fibrin,

containing a quantity of iron. The latter physiologist states that the opinion of Brande, that the amount of iron in hæmatosin is not greater than in serum and other animal substances, has been refuted by Berzelius and Engelhart. The iron is not an accidental ingredient obtained from the food; for iron has been found in the blood of a new-born animal that has never even sucked. According to Berzelius the colouring matter of the blood contains a quantity of iron corresponding to somewhat more than a half per cent. its weight of metallic iron, and he thinks it most probable that the iron exists in the blood in the metallic state, and not as an oxide.

953. By carefully watching the development of the chick in the incubated egg, the first formation of the red particles can be distinctly seen. The blood in the new being, which is elaborated before the existence of the vessels that are to contain it, is formed from the substance of the germ or from that of the germinal membrane, and is augmented by the blood of the egg, which is the substance of the yolk. First, a number of granules are produced from the substance of the yolk. These subsequently lose their granular appearance, and become translucent. On the translucent ring is produced the nucleus of the blood corpuscles. When completely formed, the blood corpuscles of the bird, as of all the animals below the bird in the scale of organization.

are of an elliptical figure, and quite flat (fig. CXII. 4, 5); but when first produced they are rounded globules, not flat, and they gradually assume their proper and permanent form; it is only on the sixth day of incubation that they begin to be elliptical, by the ninth day they are all elliptical (fig. CXII. 4, 5).

954. The substance of the fluid yolk is thus changed into blood without the action of any special organ; for, as yet, no organs such as liver, spleen, or lungs, exist. When the formation of the blood has arrived at a certain point, it begins to be in motion. The blood is seen to be in motion before the heart can be observed to beat. The germinal membrane arising out of the enlarged germinal disk soon exhibits a thin upper layer (serous membrane) and a thicker under layer (mucous membrane). There is also formed in the middle of the germinal membrane around the appearing trace of the embryo a translucent space, the *area pellucida*. The exterior of the germinal membrane remains opaque, and this opaque portion becomes divided by a definite boundary into an external and internal annular space in from sixteen to twenty hours. This separation encloses one part of the opaque portion of the germinal membrane, which surrounds the interior or translucent space of the germinal membrane, and is termed *area vasculosa*, because the blood and vessels form the inner half of this space.

955. As far as the area vasculosa extends, a granular layer is presented between the two layers of the germinal membrane, which soon divides into numerous granular isolated particles with translucent intervals, in which the blood collects, first in the form of a yellowish, and then of a reddish fluid; first distinctly in the periphery of the area vasculosa, from which it is seen to flow towards the heart before the heart beats.

956. The blood exerts its vivifying influence chiefly by the red particles. If an animal be bled to fainting, and pure serum be injected into its vessels, re-animation does not take place; but if the blood of another animal of the same species be injected, the animal which was apparently dead acquires new life at every stroke.

957. The fibrin may be removed from the blood without injuring the red particles. If the fibrin be abstracted, and a mixture of the red particles and the serum be brought to a proper temperature, and injected into the veins of an animal bled to fainting, re-animation is effected.

958. If the blood of an animal of another species be injected whose red particles are of the same form, but of a different size, re-animation is indeed effected, but the restoration is imperfect; the organic functions are oppressed, and languish, and death takes place generally within the sixth day. The same effects follow, *if* a mixture of serum and

red particles of the blood of a different species be injected.

959. If blood with circular particles be injected into the vessels of an animal whose blood corpuscles are elliptical, the most violent effects are instantly produced; such blood acts upon the nervous system like the strongest poisons; and death usually follows with extreme rapidity after the injection of a very small quantity. Thus, if a few drops of the blood of the sheep be injected into the vessels of the bird, the bird is killed instantaneously. It is very remarkable, that the blood of the mammalia should be thus fatal to the bird. The effect cannot be dependent on any mechanical principle. The injection of a fluid with particles, the diameter of which is greater than that of the capillary blood-vessels would of course destroy life by stopping the circulation; but the blood corpuscles of the mammalia are much smaller than those of the bird; yet the pigeon is killed by a few drops of mammiferous blood; and the blood of the fish is rapidly fatal to all the mammalia as well as to birds.

960. It is manifest, both from observation and experiment, that arterial blood is far more necessary to the support of the animal than of the organic life. When in asphyxia the communication of atmospheric air with the lungs is suspended, the functions of the brain are abolished;

sensibility and voluntary motion are lost the moment venous blood circulates in the arteries of the brain. It has been shown (476), that if this state continue, the animal life is destroyed in a minute and a half; but that the organic life is not extinguished for many minutes, and sometimes not even for several hours.

961. It sometimes happens that the communication between the pulmonary artery and the arota, and between the right and left auricle, which naturally exist in the fœtus, is continued after birth. In persons having this state of the circulation, called ceruleans, some portion of venous blood is always mixed with arterial blood. In this case the various processes of secretion and nutrition, the entire circle of organic functions, are but little disturbed; while the animal functions are deranged in a remarkable degree. The mind is weak and inactive, and the muscular power is so feeble, that the least exertion produces a sense of suffocation; and, if the muscular effort be continued, occasions fainting, and even suspended animation.

962. But while venous blood is in no case capable of supporting sensation and voluntary motion, there are decided cases in which secretion is effected, at least in part, from venous blood, as the bile from the venous blood that circulates through the liver in man and all the mammalia,

and the urine which is formed from venous blood in some of the lower orders of animals.

963. The proper nutritive fluid of the human body is directly formed from chyle, lymph, and venous blood; that is, partly from new matter introduced into the system from the external world, and partly from matter which has already formed a constituent part of the body. The new matter, the white chyle, is prepared partly by the action of the digestive fluids upon the food, and partly by the addition to the digested food of highly animalized substances, endowed with assimilative properties, by which the product is progressively approximated to the chemical composition of the blood. The old matter consists partly of the clear lymph, contained in the lymph vessels, and derived from the interior of the organized parts, particles which have already formed an integrant portion of the tissues and organs; and partly of the dark venous blood, the residue of the proper nutritive fluid, after the latter has yielded to the system the new matter required by it, and has given off from the system its superfluous and noxious particles.

964. In the duodenum and jejunum the new matter, the chyle, contains albumen; but it is without coagulable fibrin: it acquires fibrin in the lymph vessels on its way to the veins.

965. In the chyle globules appear; but the

chyle corpuscles are white, are without an external envelop, are comparatively few in number, are somewhat more than half the size of the blood corpuscles, and, like the nuclei of the latter, are insoluble in water.

966. The fatty or oleaginous matter contained in the chyle is in a free state, not intimately combined.

967. The chyle is alkaline, but is much less alkaline than the blood; and the iron contained in the chyle is much less intimately combined than it is in the blood.

968. Lymph contains in solution more animal matter than chyle, and the white globules are more abundant in lymph. But though lymph contain in solution more albumen and fibrin than chyle, it is not so richly loaded with these substances as blood. Still, however, the solution of albumen and fibrin in lymph approximates lymph so closely to the blood, that the lymph very much resembles the clear liquor sanguinis of which the blood consists when the red particles are abstracted from it. The colourless liquor sanguinis is the lymph of the blood. Lymph is blood without red particles; and blood, lymph with red particles.

969. The chyle is transmitted into the lymph-vessels to mingle with the lymph before it flows into the veins to mingle with the blood.

970. The commingled fluids, chyle and lymph,

pass into the blood very slowly, drop by drop. The regulation of the rapidity of the admixture seems to be the chief office of the valve placed at the termination of the thoracic duct. When the operation is observed in a living animal, it is seen that this valve prevents the new matter from flowing into the blood in a full stream. If in a dog of ordinary size that has recently eaten as much animal food as it chose, the thoracic duct be opened in the neck, the dog being alive, there will flow from the duct about half an ounce of fluid in five minutes (831); yet when this fluid reaches the termination of the duct only a few inches further on, it flows into the vein only drop by drop, at considerable intervals. One great object of pouring the chyle and lymph into the venous system so close to the heart (fig. CLXXVIII.), and of causing the commingled fluid to pass under the action of that powerful engine before it is transmitted to the lungs, seems to be, by the agitation to which it is subjected in the right auricle and ventricle to accomplish the most perfect admixture possible between the particles of the chyle and lymph and the red particles of the venous blood; an object which would be counteracted by the too rapid entrance into the current of the circulation of the new and as yet imperfectly assimilated matter.

971. After their due admixture by the powerful action of the engine that works the circulation, the commingled fluids are transmitted by the right heart to the lungs. There the watery portion of

the chyle and lymph is removed; the composition of the albumen and fibrin is completed, these substances being changed from a weak and loose into a strong and concentrated state; the solid particles are increased in number, augmented in size, and changed from a white into a red colour; carbon is given off; oxygen is absorbed; azote is alternately inhaled and exhaled; and the ultimate result is, that the three fluids—chyle, lymph, and venous blood—are converted into one homogeneous fluid, arterial blood, the proper nutrient fluid.

972. The particles of the chyle and lymph, on mingling with the blood, are scattered through the mass, and become invisible, being apparently lost among the innumerable red corpuscles; but it is not probable that the chyle is immediately converted into blood. If the coagulation of the blood be retarded by the addition of a small portion of the carbonate of potass, the red particles gradually sink some lines below the level of the fluid; and the supernatent liquid is whitish, evidently from the chylous globules mingled with the blood. In ordinary coagulation, the chyle globules are included among the immense number of the red particles of the coagulum, and are thus indistinguishable; but there is reason to believe that the chyle is not converted into blood under at least from ten to twelve hours; it is certain, that in that space of time after the completion of digestion, the serum of the blood is frequently seen to be

milk-white, from the quantity of unassimilated chyle still contained in it.

973. How the red colour of the blood is obtained, and whence the capsules of the red particles are derived, if these bodies really possess an external envelop, is wholly unknown. But it has been shown (953 and 955) that in incubation the blood is formed from the substance of the fluid yolk, without the action of any special organ; that at the period when the blood is first generated, no such organs as appear to influence the production of the blood in the adult are in existence; it is, therefore, reasonable to infer that the formation of blood in the adult may not be so dependent on the action of special organs as is commonly supposed; and that the formation of blood from chyle, of blood corpuscles from chyle corpuscles, may take place at all periods of life under the influence of the same general vital conditions as it does in the incubated egg.

974. What change the matter of the blood undergoes by respiration, whether it acquire something without which it is incapable of maintaining life, or part with something the presence of which is incompatible with life, is equally unknown. We only know that the blood, during respiration, changes its colour; but of the nature of the change produced upon its substance we are wholly iguo rant. In the present state of our knowledge, the ultimate fact is, that without the change wrought

upon the blood by respiration, the blood is incapable of maintaining life; in fact, no proper nutrient fluid is formed.

975. Once formed, the conservation of the proper proportions of the composition of the blood is effected by the excretory processes already described; by the removal of its superfluous water by the lungs, skin, and kidneys; by the removal of its superfluous carbon, azote, and oxygen by the lungs, liver, and kidneys; by the removal of saline and mineral matters chiefly by the kidneys; and finally by the instantaneous removal of products of decomposition formed in the course of the organic actions, chiefly, it would appear, by the kidneys.

976. Once formed, and duly concentrated and purified, the blood is sent out by the left heart to the system. Driven by the heart through the main trunks and branches of the aorta, the blood ultimately reaches the capillary arteries, which do not divide and subdivide indefinitely, but ultimately reach a point beyond which they no longer diminish in size. Not all of the same magnitude, some are large enough to admit of three or four of the red particles of the blood abreast; the diameter of others is only sufficient to admit of two or even of one; others are capable of transmitting only the clear and transparent liquor sanguinis; while in many cases the membranous tunics of the capillaries wholly disappear; the blood no longer flows in actual vessels, but is contained in the sub-

stance of the tissues in channels which it forms in them for itself (304).

977. Under the microscope, says Müller, the blood corpuscles are seen distinctly pouring from the smallest ramifying arteries into vessels which grow no smaller. After leaving these, they again assemble in the origins of veins formed in collected branches. The blood corpuscles flow in the finest capillaries, one after another, and often interruptedly. They are colourless when they flow singly; accumulated more thickly, they appear yellow, and in still greater quantity, yellowish red or red. In animals that have lost their strength, the globules flow without stoppage: when the animal is weak and the motion is retarded, the globules move by starts; they move on, but go more rapidly by fits. In a still weaker animal they only advance during the impulse of the heart, and then fall back a little. When several arterial currents unite in an anastomosis, one current always predominates and traverses the anastomosis alone, to mingle its blood in the other currents. Thus the currents meet and divide in the reticulate capillaries till all are collected again in veins. Sometimes the direction of the current changes, when another current becomes stronger, and the previous leader weaker, according to the pressure exerted on the part.

978. While the blood is thus traversing the capillaries, its colour changes from a bright scarlet

to a dark red. This change in the colour of the blood is the certain sign that particles have been abstracted from the circulating mass, and have been applied to the formation and support of the fluid and solid parts through which the stream is flowing. Some physiologists have satisfied themselves that they have seen the actual escape of particles from the circulating current; that they have witnessed the immediate combination of those particles with the substance of the tissues, and even that they have beheld other particles quitting the tissues and mingling with the flowing blood. Other physiologists doubt whether the most patient observation, aided by the most skilful management of the best glasses, can ever have rendered such phenomena matters of sense. "I imagined," says Müller, "at an early period, that I had seen something like this in the setting circulation; but by prolonging the observation I saw the globules move on if the current continued."

979. But whether the human eye have ever actually seen or not an interchange of particles between the blood and the tissues, it is absolutely certain that such an interchange does take place. For,—

1. Indubitable evidence has been stated (786, *et seq.*) of continual absorption from all parts of the body, yet there is no loss of substance; there must therefore of necessity be a proportionate deposition.

2. Equal evidence has been adduced (688)

that constant additions are made to the blood through the organs of digestion, yet the quantity of the blood in the body does not progressively and permanently increase; it follows that a quantity must be abstracted from the blood proportionate to the quantity added to it.

3. The human germ, from a scarcely visible point, by the successive additions of new matter progressively acquires the bulk of the adult man.

4. Organs whose special office it is to abstract particles from the blood for the elaboration of specific secretions consist almost entirely of congeries of blood-vessels. The agents are multiplied in proportion to the extent of the labour assigned them.

5. Growth, which is merely excess of deposition above absorption, is active in proportion to the quantity of blood which circulates through the growing part in a given time. The blood-vessels of a growing part increase in number and augment in size is proportion to the rapidity of the growth. In morbid growth, it is sometimes sufficient to stop the process merely to tie the main trunks of the arteries distributed to the part.

980. By every organ and every tissue; by the membrane, the muscle, the bone; by the brain, the heart, the liver, the lungs, particles are abstracted from the countless streams that bathe them, or that flow through them. In every case in which particles are thus abstracted by a tissue the following phenomena take place:—

1. Only those constituents of the blood are abstracted by the tissue which are of the same chemical nature as its own.

2. The constituents of the blood abstracted by a tissue, identical in chemical composition with its own, are immediately incorporated into its substance.

3. The constituents of the blood abstracted by a tissue, as they are incorporated into its substance, are not disposed fortuitously, but are arranged according to the specific organization of the tissue, and thus receive its own peculiar structure.

4. The constituents of the blood which thus receive the peculiar organization and structure of the tissue by which they are appropriated, acquire all its peculiar vital endowments.

981. It is manifest, then, that the tissues assimilate the blood just as the digestive fluids assimilate the aliment. And this is nutrition, the assimilation of the blood by the tissues and organs. Digestion is the conversion of the food into blood; nutrition is the conversion of blood into living fluids and solids.

982. For the reasons assigned (757 and 758), it is probable that the living fluids and solids, formed from the blood by the act of nutrition, are not generated at the parts of the body where they appear, but that, pre-existing in the blood, they are merely evolved at those parts. Hence the variety and complexity of the processes for the elaboration of the blood which have been described,

and all of which appear to be indispensable to bring the blood to a proper state of purity and strength. The great effort of the system is put forth in effecting the constitution of the blood. When the blood is once formed, all the rest of the work appears to be easy; because, before it reaches any part of the organization which it is destined to support, the blood is already adapted, mechanically, chemically, and vitally, to afford that support. Still since there are cases, as in the production of gelatin, in which the substance does not appear to be pre-existent in the blood, we are under the necessity of supposing that a material change is effected in the constituents of the vital fluid at the time and place of their escape from the circulation.

983. How the constituents of the blood escape from the circulation and incorporate themselves with the substance of the tissues there can be no difficulty in conceiving, wherever the capillaries terminate in membraneless canals, channels worked out for the reception of the nutrient stream by the force of the current itself; and in every case in which the capillaries, retaining their membranous tunics, remain true and proper vessels, their contents escape through their delicate walls by the process of endosmose (803), for which their structure appears to be admirably adapted

984. But in the capillary vessels there exists only blood. Universally and invariably before the blood passes from under the influence of the

capillary vessels it has ceased to be blood. Arterial blood is conveyed by the carotid artery to the brain; but the cerebral arteries do not deposit blood, but brain. Arterial blood is conveyed by the capillary arteries to bone; but the osseous capillaries do not deposit blood, but bone. Arterial blood is conveyed by the muscular arteries to muscle, but the muscular capillaries do not deposit blood but muscle. The blood conveyed by the capillaries of brain, bone, and muscle is the same; all comes alike from the systemic heart, and is alike conveyed to all tissues; yet in the one it becomes brain, in the other bone, and in the third muscle. Out of one and the same fluid are manufactured cuticle, and membrane, and muscle, and brain, and bone; the tears, the wax, the fat, the saliva, the gastric juice, the milk, the bile, all the fluids, and all the solids of the body (310).

985. These phenomena are wholly inexplicable on any known mechanical principles. It is equally impossible to refer them to mere chemical agency, or to any properties of dead matter. We are therefore under the necessity of referring them to a principle which, for the sake of distinguishing it from anything mechanical or chemical, we term vital. As the actions which take place between the integrant particles of bodies, giving rise to chemical phenomena, are referred to one general principle, termed chemical affinity, so the actions which take place in living bodies, giving rise to

vital phenomena, may be referred to one general principal, termed vital affinity. The term explains nothing, it is true, it merely expresses the general fact; but still it is convenient to have a term for the expression of the fact. The property itself will ever remain an ultimate fact in physiology, however exactly the limits of its agency, and the laws according to which it modifies the mechanical and chemical relations of the substances subjected to its influence, may hereafter be ascertained; just as chemical affinity will ever be an ultimate fact in physics, whatever discoveries may yet be made of the extent of its agency and of the conditions on which its action depends.

986. It is then an ascertained fact, that there exists between the blood and the tissues a mutual reaction, not of a physical, but of a vital nature, in which the blood takes as active a part as the tissue, and the tissue as the blood; the blood exerting a vital attraction on the tissue, and the tissue on the blood. We only express this ultimate fact when we say (and this is all we can do) that in every part of the body, by virtue of a vital affinity, the tissue attracts from the blood the molecules of matter appropriate to its chemical composition, and the blood attracts from the tissue the particles which, having served their purpose there, are destined to other uses in the economy; or, if wholly useless, are absorbed into the current of the circulation to be expelled from the system.

987. We can see how the particles of matter which are attracted by the tissue from the blood are so deposited and disposed that the tissue always preserves its own shape, bulk, and relation to the surrounding tissues. This definite arrangement is the result of an action which has been already stated to be proper to the absorbent vessels. Previously to the deposition of a new particle of matter by a capillary, an old particle is removed by an absorbent, either a lymphatic or a vein. In removing the old matter, the absorbent forms a mould into which the capillary deposits the new molecules; and the form of every tissue and organ depends on the kind of mould formed for the reception of its nutrient matter by the absorbent vessel. The absorbents are thus the architects of the system; and the capillaries are both chemists which form the rough material employed in the structure, and masons which deposit and arrange it. The conjoint action of both sets of vessels is necessary to the formation of the simplest tissue; and it is by their united labour that the compound organs are built up out of the simple tissues.

988. It is conjectured that the immediate living agents by which this vital attraction is exerted between the blood and the tissues are the organic nerves. These nerves consist of two sets, those which enter as constituents into the tissues and those which accompany the capillaries. It has been shown (304), that while the membranous

tunics of the capillaries diminish, the nervous filaments distributed to them increase; that the smaller and thinner the capillaries the greater the proportionate quantity of their nervous matter; and that this is most remarkably the case in organs of the greatest irritability. It is conceived that the capillaries, in consequence of the nervous structure which thus envelops them, exert upon the fluid which is flowing through them an influence perfectly analogous to that of the secreting organ, in consequence of which similar particles are abstracted from the blood as those which compose the tissue in which the operation takes place.

989. It is further conjectured that the physical agent by which this action upon the blood is effected is the galvanic fluid. Dutrochet believes that he has actually formed muscular fibre from albumen by galvanism. He considers the red particles of the blood as pairs of electrical plates, and thinks that the nucleus is electronegative, and the capsule electropositive. Müller has repeated and critically examined the interesting experiments of Dutrochet; and while he arrives in many essential points at different results, expresses the highest admiration of the ingenious manner in which this philosopher has sought to solve a great problem. "If," says Müller, "a drop of an aqueous solution of the yolk of egg (in which very small microscopic globules are suspended) be galvanised, the currents discovered by Dutrochet will be observed. The wave,

proceeding from the copper or negative pole, in which the alkali of the decomposed salt accumulates, is transparent, from the solution of albumen by the alkali. The wave, proceeding from the positive or zinc pole, particularly in its circumference, is opaque, and white from the acid it contains. Both waves encounter, and exactly in the line of contact a linear coagulum is immediately produced, which assumes the form of the line of contact, and is curled at times as the edges of the waves are meeting. The meeting of both waves takes place with a lively motion, in the line of contact, when the deposition of coagulum takes place; but as soon as the deposition of coagulum has occurred, all is tranquil, and not the least trace of motion is observed. It is therefore inconceivable how an observer of the first rank, like Dutrochet, can pronounce this coagulated albumen contractile muscular fibre, generated by galvanism; it is nothing but coagulated albumen. This coagulum, besides, like the albumen which is deposited by galvanism round the zinc pole, has no consistence, but is composed of globules easily separated by stirring, and only precipitated in the line where the two waves meet without cohesion."

990. But though science has not yet succeeded in ascertaining with certainty the physical agency to which the ultimate changes that take place in organized matter are to be referred, there cannot be a question that they are dependent on physical

agents; and the legitimate object of scientific inquiry is to discover what those agents are, and to ascertain the modifications they undergo by those vital affinities to the influence of which they are subjected.

991. The discoveries which science has already made relative to the influence of certain physical agents on particular organs, and to the influence of the whole circle of physical agents on the whole living economy, have added not a little to human power over human health and disease. But these agents also exert an influence scarcely less momentous on the entire apparatus and action of the animal life, so inseparably linked with the organic. An account will therefore be next given of the structure and function of the nervous and muscular systems. The exposition of these systems, which will be as brief as possible, will be followed by a full account of the action of physical agents on the whole of this complex and wonderful organization. The detail of the ascertained phenomena will have a strict reference to the development of the physical and mental powers of the human being, and thereby a close and practical application will be attempted of physiology to the production and preservation of health.

THE END.

Lightning Source UK Ltd.
Milton Keynes UK
UKHW020609110119
335177UK00005B/334/P